重庆市教育科学"十三五"规划(课题编号:2017 – GX – 181)

WCF 技术探析

邹劲松 著

哈尔滨工程大学出版社
Harbin Engineering University Press

内容简介

WCF 技术是基于 SOA 的编程思想,它从统一性、互操作性、安全与可靠性、兼容性等特性上体现其强大的生命力。本书内容共分为七个部分,包括 WCF 基础知识,服务契约的定义及实现过程,数据协定的定义及实现过程,WCF 服务应用程序的分布式事务,元数据、客户端的服务调用原理及过程,以及 WCF 的常见服务承载方式等。

本书适用于计算机专业的高职或本科学生,也面向基于 NET 平台的 C#程序开发人员、.NET 架构师、SOA 架构师或企业系统集成架构师及所有 WCF 爱好者。

图书在版编目(CIP)数据

WCF 技术探析/邹劲松著. —哈尔滨:哈尔滨工程大学出版社,2018.7
ISBN 978 – 7 – 5661 – 1959 – 9

Ⅰ. ①W… Ⅱ. ①邹… Ⅲ. ①网络服务器 – 程序设计 Ⅳ. ①TP368.5

中国版本图书馆 CIP 数据核字(2018)第 117310 号

选题策划	王洪菲
责任编辑	雷 霞
封面设计	李海波

出版发行	哈尔滨工程大学出版社
社　　址	哈尔滨市南岗区南通大街 145 号
邮政编码	150001
发行电话	0451 – 82519328
传　　真	0451 – 82519699
经　　销	新华书店
印　　刷	黑龙江龙江传媒有限责任公司
开　　本	787 mm × 1 092 mm　1/16
印　　张	12.25
字　　数	268 千字
版　　次	2018 年 7 月第 1 版
印　　次	2018 年 7 月第 1 次印刷
定　　价	49.80 元

http://www.hrbeupress.com
E-mail:heupress@ hrbeu.edu.cn

前 言

Windows Communication Foundation(WCF)是由微软开发的一系列支持数据通信的应用程序框架,整合了原有的 Windows 通信的 .NET Remoting、WebService、Socket 的机制,并融合了 HTTP 和 FTP 等相关技术。

WCF 3.0 自 2006 年诞生以来,受到很多编程爱好者的追捧,同时也被许多软件公司用于商业软件开发,取得了不俗的市场份额。WCF 作为 .NET 平台的重要组成部分,为开发人员提供构建跨平台、安全、可靠和支持事务处理的企业级互联应用解决方案。

WCF 的优势:(1)统一性。它是多种技术的整合,但仍然可以像 .NET 一样面向对象代码编写,因为它采用托管代码编写方式。(2)互操作性。它采用 SOAP 通信机制,保证了系统之间的互操作性,可以跨进程、跨机制,甚至跨平台通信,只要它支持 Web Service。(3)安全与可靠性。因为它包含多种协议,故有多种协议的安全机制,尤其是 SOAP 的。(4)兼容性。它可以在新、旧平台上使用。

本书是著者针对高等职业技术教育软件开发专业,根据自己的教学心得、开发经验,面向广大高职学生着力提升软件编程能力而开发的基于 SOA 编程思想的 WCF 技术专著。

本书也适合基于 .NET 平台的 C#程序开发人员、软件测试工程师、.NET 架构师、SOA 架构师或企业系统集成架构师及所有 WCF 爱好者阅读。

本书共 7 章,其主要内容如下:

第 1 章介绍 WCF 的基础知识。包括 WCF 常见术语、服务的设计和实现、服务端配置文件的 WCF 服务配置、客户端服务的访问和配置客户端的行为,其目的是让读者对 WCF 的框架体系、实现原理及过程等有一个清晰明确的认识。

第 2 章介绍服务契约的定义、实现过程。包括单向、双工、错误协定、会话、流等几种常见模式,其目的是让读者掌握服务契约的各种模式的实现原理、代码实现。

第 3 章介绍数据协定的定义、实现过程。包括数据协定基本概念的介绍、创建类或结构来实现基本数据协定、可序列化类型、数据成员顺序、数据协定已知类型和数据协定序列化程序等。

第 4 章介绍构建 WCF 服务应用程序的分布式事务。包括事务基本概念理解、ServiceModel 事务特性、ServiceModel 事务配置、启用事务流和创建事务性服务等。

第 5 章介绍元数据在 WCF 编程中的作用。包括元数据体系结构、导出及导入元数据、发布元数据、检索元数据和使用元数据等相关知识点。

第 6 章介绍客户端的服务调用原理及过程。包括 WCF 客户端体系结构、使用单向和请求–答复协定访问 WCF 服务、使用双工协定访问服务、使用 ChannelFactory、以异步方式调用 WCF 服务操作、使用通道工厂以异步方式调用操作、创建通道工厂并用它创建和管理通道、客户端配置等。

第 7 章介绍 WCF 服务的几种常见承载方式。包括 IIS 承载、WAS 承载和 Windows 服务应用程序中承载等。

阅读本书应具备基于.NET 平台的程序开发语言,主要以 C#为主,还有.NET 架构基础知识、网络基本知识、进程、线程与应用程序域、数字墨迹与动态绘图基础、数据流与数据的加密、解密和 ASP. NET 等。

本书在撰写过程中得到了重庆市教育科学"十三五"规划 2017 年度重点无经费课题"高职供给侧改革下软件技术专业现代学徒制育人模式的实践探索"(课题编号:2017 – GX – 181。主持人:邹劲松),重庆市高等教育学会 2017—2018 年度高等教育科学研究课题"高职供给侧改革视域下水务管理信息化课程体系重构研究与实践"(项目编号:CQGJ17046A。主持人:邹劲松),重庆市高等职业技术教育研究会"十三五"高等职业教育科学研究规划课题(项目编号:GY171002。主持人:邹劲松)和中国职业技术教育教学工作委员会、教材工作委员会 2017—2018 年度教学改革与教材建设课题"智慧水务视阈下水务管理人才培养模式改革研究与实践"(项目编号:1711065。主持人:邹劲松)等的资助。本书部分内容得到马骏、徐雷等专家和互联网上 WCF 爱好者的大力支持,在此表示衷心的感谢。

最后,对哈尔滨工程大学出版社的领导和编辑等相关工作人员在本书撰写和编辑过程中付出的辛勤劳动深表感谢。

由于著者水平有限,书中难免存在不妥或错误之处,敬请专家、读者批评指正。

著 者

2018 年 3 月

目　　录

第1章　WCF 基础 1
　1.1　基本编程生命周期 1
　1.2　基本概念概述 1
　1.3　设计和实现服务 7
　1.4　服务配置 15
　1.5　客户端生成 25

第2章　服务契约 39
　2.1　单向 39
　2.2　双工 41
　2.3　错误协定 46
　2.4　会话 49
　2.5　流 51

第3章　数据协定 57
　3.1　数据协定基本知识 57
　3.2　可序列化类型 61
　3.3　数据成员顺序 62
　3.4　数据协定已知类型 64
　3.5　数据协定序列化程序 71

第4章　事务 84
　4.1　事务概述 84
　4.2　事务性支持 85

第5章　元数据 98
　5.1　元数据体系结构概述 98
　5.2　导出和导入元数据 102
　5.3　发布元数据 114
　5.4　检索元数据 124
　5.5　使用元数据 127

第 6 章　客户端 ··· 139
　6.1　WCF 客户端体系结构 ·· 139
　6.2　WCF 客户端访问服务 ·· 142
　6.3　WCF 客户端配置 ··· 162
第 7 章　承载 ··· 166
　7.1　IIS 承载 ··· 166
　7.2　WAS 承载 ·· 174
　7.3　Windows 服务应用程序中承载 ······································ 185
　7.4　托管应用程序中承载 ··· 188
参考文献 ·· 189

第 1 章 WCF 基础

1.1 基本编程生命周期

Windows Communication Foundation（WCF）可让应用程序通报它们是在同一台计算机上、分布在 Internet 上还是在不同的应用程序平台上。

要执行的基本任务依次为：

（1）定义服务协定。服务协定指定服务的签名、服务交换的数据和其他协定要求的数据。

（2）实现协定。若要实现服务协定，请创建实现协定的类并指定运行时应具有的自定义行为。

（3）通过指定终结点和其他行为信息来配置服务。

（4）承载服务。

（5）生成客户端应用程序。

尽管本节中的主题遵循此顺序，但是一些方案并不会从头开始。例如，如果想要为预先存在的服务生成客户端，则从步骤(5)开始。或者，如果用户是在生成其他人要使用的服务，则可以跳过步骤(5)。

1.2 基本概念概述

WCF 就是专门用于服务定制、发布、运行，以及消息传递和处理的一组专门类的集合，也就是所谓的"类库"。这些类通过一定方式被组织起来，共同协作，并为开发者提供了一个统一的编程模式。WCF 之所以特殊，是在于它所应对的场景与普通的.NET 类库不同，它主要用于处理进程间乃至机器之间消息的传递与处理，同时它引入了 SOA 的设计思想，以服务的方式公布并运行，以方便客户端跨进程和机器对服务进行调用。实际上，WCF 就是微软对于分布式处理的编程技术的集大成者，它将 DCOM、Remoting、Web Service、WSE、MSMQ 集成在一起，从而降低了分布式系统开发者的学习曲线，并统一了开发标准。

1. 消息和终结点

WCF 建立在基于消息的通信这一概念基础之上,可以建模为消息(如 HTTP 请求或消息队列(也称为 MSMQ)消息)的任何内容都可以在编程模型中按照统一方式进行表示。这样,就可以在不同传输机制间提供一个统一的 API。

模型区分客户端(应用程序启动的通信)和服务,即服务等待客户端的服务请求和响应其服务的通信过程。单个应用程序既可以充当客户端,也可以充当服务。

消息在终结点之间发送。终结点可以发送或接收与其所有相关的信息。服务端公开一个或多个应用程序终结点,而客户端生成一个与服务端终结点相兼容的终结点。

终结点以基于标准的方式进行阐述,着重分析消息的发送与接收过程及原理。服务端向客户端公开消息,使其可以处理 WCF 客户端生成的元数据和通信堆栈。

2. 通信协议

一个必需的通信堆栈元素是传输协议,可以使用常用传输协议(如 HTTP 和 TCP)通过 Intranet 和 Internet 发送消息,也可以使用其他支持与消息队列应用程序和对等网络网格上的节点进行通信的传输协议。使用 WCF 的内置扩展点可以添加更多传输机制。

通信堆栈中的另一个必要元素是指定如何将任意给定消息进行格式化的编码。WCF 提供了以下类型编码:

◇ 文本编码,一种可互操作的编码;
◇ 消息传输优化机制(MTOM)编码,是一种可互操作的方法,用于高效地将非结构化二进制数据发送到服务或从服务接收这些数据;
◇ 用于实现高效传输的二进制编码。

使用 WCF 的内置扩展点可以添加更多编码机制(如压缩编码)。

3. 消息模式

WCF 支持多种消息模式,包括请求–答复、单向和双工通信。不同传输协议支持不同的消息模式,因而会影响它们所支持的交互类型。WCF API 和运行库还能帮助用户安全而可靠地发送消息。

4. WCF 术语

(1)消息

消息是一个独立的数据单元,它可能由几个部分组成,含消息正文和消息头。

(2)服务

服务是一个构造,它公开一个或多个终结点,其中每个终结点都公开一个或多个服务操作。

(3)终结点

终结点是用来发送或接收消息(或同时执行这两种操作)的构造。终结点包括一个定义消息可以发送到的目的地的位置(地址)、一个描述消息应如何发送的通信机制规范(绑定),以及对可以在该位置发送或接收(或同时执行这两种操作)的一组消息的定义(服务协定,用于描述可以发送哪些消息)。

WCF 服务作为一个终结点集合对外公开。

(4)应用程序终结点

一个终结点,由应用程序公开并对应于该应用程序实现的服务协定。

(5)基础结构终结点

一个终结点,由基础结构公开,以便实现与服务协定无关的服务需要或提供的功能。例如服务可能拥有一个提供元数据信息的基础结构终结点。

(6)地址

地址用于指定接收消息的位置。地址以统一资源标识符(URI)的形式指定。URI 架构部分指定用于到达地址的传输机制,如 HTTP 和 TCP。URI 的层次结构部分包含一个唯一的位置,其格式取决于传输机制。

使用终结点地址可以为服务中的每个终结点创建唯一的终结点地址,或者在某些条件下在终结点之间共享一个地址。下面的示例演示了一个将 HTTPS 协议和一个非默认端口结合使用的地址:

HTTPS://cohowinery:8005/ServiceModelSamples/CalculatorService

(7)绑定

绑定定义终结点与外界进行通信的方式。它由一组称为绑定元素的要素构造而成,这些元素"堆叠"在一起以形成通信基础结构。绑定最起码应定义传输协议(如 HTTP 或 TCP)和所使用的编码(如文本或二进制)。绑定可以包含指定详细信息(例如用于保护消息的安全机制或终结点所使用的消息模式)的绑定元素。

(8)绑定元素

绑定元素表示绑定的特定部分,如传输协议、编码、基础结构级协议(如 WS – Reliable Messaging)的实现以及通信堆栈的其他任何要素。

(9)行为

行为是控制服务、终结点、特定操作或客户端的各个运行方面的要素。行为按照范围进行分组:常见行为在全局范围内影响所有终结点;服务行为仅影响与服务相关的方面;终结点行为仅影响与终结点相关的属性;操作级行为影响特定操作。例如有一种服务行为是遏制,它指定当过多的消息可能超出服务的处理能力时,服务应该如何反应。另一方面,终结点行为仅控制与终结点的相关方面,如查找安全凭据的方式和位置。

（10）系统提供的绑定

WCF 包含许多系统提供的绑定。这些绑定是针对特定方案进行优化的绑定元素的集合。例如 WSHttpBinding 是为了与各种 WS* 规范服务进行互操作而专门设计的。通过仅提供那些可以正确应用于特定方案的选项，这些预定义的绑定可以节省时间。如果预定义的绑定不能满足其要求，则可以创建其自身的自定义绑定。

（11）配置与编码

可以通过代码编写、配置或将两者结合在一起对应用程序进行控制。配置的优点在于，它使非开发人员（如网络管理员）可以在代码编写完成后直接对客户端和服务参数进行设置，而不必重新进行编译。使用配置不仅可以设置值（如终结点地址），还可以通过添加终结点、绑定和行为来实施进一步的控制。通过代码编写，开发人员可以保持对服务或客户端的所有组件的严格控制，而且可以对通过配置完成的所有设置进行检查，并根据需要通过代码进行重写。

（12）服务操作

服务操作是在服务的代码中定义的过程，用于实现某种操作的功能。此操作作为一个 WCF 客户端上的方法向客户端公开。该方法可以返回一个值，并可采用数量可选的参数，或是不采用任何参数且不返回任何响应。例如一个实现"Hello"的简单操作可以用作客户端存在通知，并可以开始一系列操作。

（13）服务协定

服务协定将多个相关的操作联系在一起，组成单个功能单元。协定可以定义服务级设置，如服务的命名空间、对应的回调协定以及其他此类设置。在大多数情况下，协定的定义方法是用所选的编程语言创建一个接口，然后将 ServiceContractAttribute 属性应用于该接口。通过该接口可生成实际的服务代码。

（14）操作协定

操作协定定义参数并返回操作的类型。在创建定义服务协定的接口时，可以通过将 OperationContractAttribute 属性应用于协定中包含的每个方法定义来表示一个操作协定。可以将操作建模为采用单个消息作为参数并返回单个消息，或者建模为采用一组类型作为参数并返回一个类型。在后一种情况下，系统将确定需要为该操作交换的消息的格式。

（15）消息协定

消息协定描述消息的格式。例如它会声明消息元素应包含在消息头中还是包含在消息正文中，应该对消息的何种元素应用何种级别的安全性，等等。

（16）错误协定

可以将错误协定与服务操作进行关联，以指示可能返回到调用方的错误。一个操作可以具有零个或更多个与其相关联的错误。这些错误是在编程模型中作为异常建模的 SOAP 错误。

(17)数据协定

服务使用的数据类型必须在元数据中进行描述,以使其他各方可以与该服务进行交互操作。数据类型可以在消息的任何部分使用(例如作为参数或返回类型)。如果服务仅使用简单类型,则无须显式使用数据协定。

(18)承载

服务必须承载于某个进程中。主机是控制服务的生存期的应用。服务可以是自承载的,也可以由现有的托管进程管理。

(19)自承载服务

自承载服务是在开发人员创建的进程应用程序中运行的服务。开发人员控制服务的生存期、设置服务的属性、打开服务(这会将服务设置为侦听模式)以及关闭服务。

(20)宿主进程

宿主进程是专为承载服务而设计的应用程序。这些宿主进程包括 Internet 信息服务(IIS)、Windows 激活服务(WAS)和 Windows 服务。在这些宿主方案中,由宿主控制服务的生存期。例如使用 IIS 可以设置包含服务程序集和配置文件的虚拟目录。在收到消息时,IIS 将启动服务并控制服务的生存期。

(21)实例化

每个服务都具有一个实例化模型。有三种实例化模型:"单个",在这种模型中,由单个 CLR 对象为所有客户端提供服务;"每个调用",在这种模型中,将创建一个新的 CLR 对象来处理每个客户端调用;"每个会话",在这种模型中,将创建一组 CLR 对象,并且为每个独立的会话使用一个对象。实例化模型的选择取决于应用程序需求和服务的预期使用模式。

(22)客户端应用程序

客户端应用程序是与一个或多个终结点交换消息的程序。客户端应用程序通过创建一个 WCF 客户端实例并调用该 WCF 客户端的方法来开始工作。需要注意的是,单个应用程序既可以充当客户端,也可以充当服务。

(23)通道

通道是绑定元素的具体实现。绑定表示配置,而通道是与该配置相关联的实现,因此每个绑定元素都有一个相关联的通道。通道堆叠在一起以形成绑定的具体实现:通道堆栈。

(24)WCF 客户端

WCF 客户端是一个将服务操作作为方法公开的客户端应用程序构造。任何应用程序都可以承载 WCF 客户端,包括承载服务的应用程序,因此可以创建一个包含其他服务的 WCF 客户端的服务。

WCF 客户端可以通过使用自动生成 Service Model 元数据实用工具(Svcutil.exe)并指向正在运行的服务发布的元数据。

(25) 元数据

服务的元数据描述服务的各种特征,外部实体需要了解这些特征以便与该服务进行通信。元数据可供 Service Model 元数据实用工具(Svcutil.exe)生成 WCF 客户端和客户端应用程序可用于与服务交互的伴随配置。

服务所公开的元数据包括 XML 架构文档(用于定义服务的数据协定)和 WSDL 文档(用于描述服务的方法)。

启用元数据后,WCF 通过检查服务及其终结点自动生成服务的元数据。若要发布服务的元数据,必须显式启用元数据行为。

(26) 安全性

WCF 中的安全性包括保密性(为防止窃听而进行的消息加密)、完整性(用于检测消息篡改行为的方法)、身份验证(用于验证服务器和客户端的方法)以及授权(资源访问控制)。通过利用现有安全机制(如 TLS over HTTP,也称为 HTTPS)或通过实现各种 WS - * 安全规范中的一个或多个规范,可以提供这些功能。

(27) 传输安全模式

传输安全模式指定由传输层机制(如 HTTPS)提供保密性、完整性和身份验证。在使用像 HTTPS 这样的传输协议时,此模式的优点在于性能出色,而且由于它在 Internet 上非常流行,因此很容易理解。其缺点在于,这种安全分别应用于通信路径中的每个跃点,这使得通信容易遭受"中间人"攻击。

(28) 消息安全模式

确定安全通过可能涉及一个或多个安全规范中,如提供规范名为 Web 服务的 SOAP 消息安全。每个消息都包含必要的安全机制,用于在消息传输过程中保证安全,并使接收方能够检测到篡改和对消息进行解密。从这种意义上说,安全信息包装在每个消息中,从而提供了跨多个跃点的端到端安全。由于安全信息成为消息的一部分,还有可能包含多种凭据并显示消息(这些被称为声明)。这种方法还具有这样一个优点,即消息可以通过任意传输协议(包括在其起点和目标之间的多个传输协议)安全地传送。这种方法的缺点在于所使用的加密机制较为复杂,使性能受到影响。

(29) 使用消息凭据的传输安全模式

此模式指定使用传输层来提供消息的保密性、身份验证和完整性,并且每个消息都可以包含消息接收方所要求的多个凭据(声明)。

(30) WS - *

一组不断增加的、在 WCF 中予以实现的 Web 服务(WS)规范(如 WS - Security、WS - Reliable Messaging 等)的简写。

1.3 设计和实现服务

本节介绍如何定义和实现 WCF 协定。服务协定指定终结点与外界通信的内容。更具体地说,它是有关一组特定消息的声明,这些消息被组织成基本消息交换模式(MEP),如请求—答复、单向和双工。如果说服务协定是一组在逻辑上相关的消息交换,那么服务操作就是单个消息交换。例如"Hello"操作显然必须接受一条消息(以便调用方能够发出问候),并可能返回也可能不返回一条消息。

1.3.1 概述

本部分介绍 WCF 服务的设计和实现提供高级概念性阐释。子主题提供有关具体设计和实现过程更为详细的信息。在设计和实现 WCF 应用程序之前就理解以下内容:

◇ 了解什么是服务协定、服务协定的工作原理以及如何创建服务协定;
◇ 了解运行时配置或宿主环境可能不支持的协定状态最低要求。

1. 服务协定

服务协定指定以下内容:

◇ 协定公开的操作;
◇ 针对交换的消息所进行的各种操作的签名;
◇ 这些消息的数据类型;
◇ 操作的位置;
◇ 用于支持与服务成功通信的特定协议和序列化格式。

例如,采购订单协定可能具有一个 CreateOrder 操作,该操作接受订单信息类型输入并返回成功或失败信息,包括一个订单标识符。它还可能具有一个 GetOrderStatus 操作,该操作接受一个订单标识符并返回订单状态信息。此类服务协定需要指定以下四方面:

◇ 采购订单协定由 CreateOrder 和 GetOrderStatus 操作组成;
◇ 这些操作指定了输入消息和输出消息;
◇ 这些消息可以携带的数据;
◇ 有关成功处理消息所必需的通信基础结构的分类声明,例如这些详细信息包括建立成功通信是否需要安全以及需要哪些形式的安全。

若要表达这种对其他应用程序在许多平台(包括非 Microsoft 平台)上的信息,XML 服务协定公开以表示标准 XML 格式,如 Web 服务描述语言(WSDL)和 XML 架构(XSD)等。许多平台的开发人员都可以使用此公共协定信息创建可与该服务通信的应用程序,因为开发人员理解规范的语言,且这些语言通过描述服务支持的公共形式、格式和协议,支持互操作。

协定可以用多种方式表示,虽然 WSDL 和 XSD 语言非常适合以易于理解的方式描述服务,但这些语言很难直接使用,它们仅用于描述服务,而不能描述服务协定实现,因此 WCF 应用程序使用托管属性、接口和类来定义服务的结构并实现服务。

当客户端或其他服务实施者需要时,可将托管类型中定义的协定导出作为元数据,即 WSDL 和 XSD。

结果可以得到一个简单的编程模型,可以使用公共元数据向任何客户端应用程序描述该模型。基础 SOAP 消息的详细信息、传输和安全相关信息等可以留给 WCF 处理,它可以在服务协定类型系统和 XML 类型系统之间自动执行必要的转换。

2. 面向消息

如果用户习惯于远程过程调用(RPC)样式的方法签名(其请求功能的标准形式为向某一方法传递参数,然后从对象或其他类型的代码接收返回值),则使用托管接口、类和方法模拟服务操作简单而直观。例如使用托管语言(如 Visual Basic 和 C++ COM)的程序员可以运用 RPC 样式方法(不管是使用对象还是接口)方面的知识来创建 WCF 服务协定,而不会遇到 RPC 样式分布式对象系统中固有的问题。面向服务的优势是可以实现松耦合、面向消息的编程,同时保持熟悉的 RPC 编程体验。

许多程序员感觉使用面向消息的应用程序编程接口更舒服,例如像 Microsoft MSMQ 这样的消息队列,.NET Framework 中的 System.Messaging 命名空间或在 HTTP 请求中发送非结构化 XML。

3. 了解需求的层次结构

服务协定对操作进行分组,指定消息交换模式、消息类型和消息携带的数据类型,并指示实现为了支持该协定而必须具有的运行时行为类别(例如可能要求对消息进行加密和签名)。服务协定本身并未明确指定如何满足这些要求,而只是指定这些要求必须得到满足。加密类型或消息的签名方式取决于相容服务的实现和配置。

请注意协定为添加行为而对服务协定实现和运行时配置提出特定要求的方式。为了公开某一服务以供使用而必须满足的这组要求建立于前一组要求之上。如果协定对实现提出要求,那么实现可能会对配置和绑定提出更多的要求,以便使服务能够运行。最后,主机应用程序还必须支持服务配置和绑定所添加的任何要求。

在设计、实现、配置和承载 WCF 服务应用程序时必须要记住这一添加要求的过程。例如协定可能会指定需要支持某一会话。如果是这样,用户必须配置绑定以支持该协定性需求,否则服务实现将无法正常工作。或者,如果用户的服务要求 Windows 集成身份验证并寄宿在 Internet 信息服务(IIS)中,则服务所在的 Web 应用程序必须打开 Windows 集成身份验证并关闭匿名支持。

1.3.2 设计服务协定

本部分介绍什么是服务协定、如何定义服务协定、可用的操作(以及基础消息交换的含义)、使用的数据类型以及可帮助用户设计能满足方案需求的操作的其他问题。

1. 创建服务协定

服务公开一系列操作。在 WCF 应用程序中,通过创建一个方法并使用 OperationContractAttribute 属性对其进行标记来定义操作。然后,若要创建服务协定,需要将操作组合到一起,具体方法是在使用 ServiceContractAttribute 属性标记的接口中声明这些操作,或在使用同一属性进行标记的类中定义它们。

任何不具有 OperationContractAttribute 特性的方法都不是服务操作,不能由 WCF 服务公开。

本部分介绍设计服务协定时的以下决策要点:
◇ 是否使用类或接口。
◇ 如何指定要交换的数据类型。
◇ 用户可以使用的交换模式类型。
◇ 是否可以将显式安全要求作为协定的一部分。
◇ 对操作输入和输出的限制。

2. 类和接口

类和接口都表示一组功能,因此二者都可用于定义 WCF 服务协定。但是建议用户使用接口,因为接口可以直接对服务协定建模。如果不经过实现,接口的作用只是根据特定签名对一组方法进行定义。如果实现服务协定接口,即可实现 WCF 服务。

服务协定接口具有托管接口的所有优点;服务协定接口可以扩展任何数量的其他服务协定接口。一个类可以通过实现服务协定接口来实现任意数量的服务协定。

用户可以通过更改接口实现来修改服务协定的实现,而让服务协定保持不变。您可以通过实现旧接口和新接口来确定服务的版本。老客户端连接到原始版本,而新客户端则可以连接到较新的版本。

备注

从其他服务协定接口中继承时,不能重写操作属性,例如名称或命名空间。如果试图执行该操作,应在当前服务协定中创建新操作。

不过,用户可以使用类来定义服务协定,并同时实现该协定。可以通过直接向类和类上的方法分别应用 ServiceContractAttribute 和 OperationContractAttribute 来创建服务,这种方法的优点是快速且简便。缺点是托管类不支持多个继承,因此一次只能实现一个服务协定。

此外,对类或方法签名的任何修改都将修改该服务的公开协定,这可以防止未进行修改的客户端使用用户的服务。

此时,用户应了解使用接口定义服务协定与使用类定义服务协定之间的区别。下一步将确定可在服务及其客户端之间往返传递的数据。

3. 参数和返回值

每个操作都有一个返回值和一个参数,即使它们为 void。用户可以使用局部方法将对对象的引用从一个对象传递到另一个对象,但与局部方法不同的是,服务操作不会传递对对象的引用,它们传递的只是对象的副本。

这一点很重要,这是因为参数或返回值中使用的每个类型都必须是可序列化的,换言之,该类型的对象必须能够转换为字节流,并能够从字节流转换为对象。

默认情况下,基元类型是可序列化的,.NET Framework 中的很多类型都是可序列化的。

备注

操作签名中的参数名称值是协定的一部分且区分大小写。如果要在本地使用相同的参数名称,但是要在已发布的元数据中修改名称,请参见 System.ServiceModel.MessageParameterAttribute。

4. 数据协定

面向服务的应用程序(例如 WCF 应用程序)设计为与 Microsoft 平台和非 Microsoft 平台上的最大可能数量的客户端应用程序进行互操作。为了获得最大可能的互操作性,建议您使用 DataContractAttribute 和 DataMemberAttribute 属性对您的类型进行标记,以创建数据协定。数据协定是服务协定的一部分,用于描述您的服务操作交换的数据。

数据协定是可选的样式协定:除非显式应用数据协定属性,否则不会序列化任何类型或数据成员。数据协定与托管代码的访问范围无关:可以对私有数据成员进行序列化,并将其发送到其他位置,以便可以公开访问它们。WCF 处理启用操作的功能的基础 SOAP 消息的定义,以及序列化用户的数据类型的传入和传出的消息正文。只要数据类型可序列化,就无须在设计操作时考虑基础消息交换基础结构。

尽管典型的 WCF 应用程序使用 DataContractAttribute 和 DataMemberAttribute 属性来创建用于操作的数据协定,用户仍可以使用其他序列化机制。标准 ISerializable、SerializableAttribute 和 IXmlSerializable 机制都可用于处理数据类型到基础 SOAP 消息的序列化,这些消息可将数据类型从一个应用程序带到另一个应用程序。如果用户的数据类型需要特别支持,可以采用多个序列化策略。

5. 将参数和返回值映射到消息交换

应用程序不但需要支持特定标准安全、事务和与会话相关的功能时所涉及的数据,还对其数据进行基础交换时往返传输的 SOAP 消息提供服务支持还操作。服务操作签名应指定特定的基础消息交换模式(MEP)即可以支持数据传输和操作要求的功能。当然,用户可以在 WCF 编程模型中指定三种模式:请求－答复、单向和双工消息模式。

6. 请求－答复

通过请求－答复模式,发送方(客户端应用程序)将接收与请求相关的答复。这是默认的 MEP,因为它既支持传入操作(一个或多个参数传递到该操作中),也支持将返回值传回给调用方。例如下面的 C# 代码示例演示一个基本的服务操作,即先接收一个字符串,再返回一个字符串。

```
[OperationContractAttribute]
string Hello(string greeting);
```

此操作签名指示基础消息交换的形式。如果不存在关联,则 WCF 无法确定返回值所期望的操作。

请注意,除非指定其他基础消息模式,否则即使服务操作返回 void(在 Nothing 中为 Visual Basic),也属于请求－答复消息交换。操作的结果是:除非客户端异步调用操作,否则客户端将停止处理,直到收到返回消息,即使该消息正常情况下为空时也是如此。下面的 C# 代码示例演示的操作在客户端收到空的响应消息后才返回值。

```
[OperationContractAttribute]  void Hello(string greeting);
```

如果执行操作需要很长的时间,则上面的示例会降低客户端性能和响应能力,但是即使在请求－答复操作返回 void 时,这种操作仍有优势。最明显的优势在于,响应消息中可返回 SOAP 错误,这表明可能在通信或处理中发生了一些与服务有关的错误状况。在服务协定中指定的 SOAP 错误将作为 FaultException<TDetail> 对象传递到客户端应用程序,其中类型参数是在服务协定中指定的类型。这使得将 WCF 服务的错误状况通知给客户端的过程变得很方便。

7. 单向

如果 WCF 服务应用程序的客户端不必等待操作完成,并且不处理 SOAP 错误,则该操作可以指定单向消息模式。单向操作是客户端调用操作并在 WCF 将消息写入网络后继续进行处理的操作。通常这意味着,除非在出站消息中发送的数据极其庞大,否则客户端几乎立即继续运行(除非发送数据时出错)。此种类型的消息交换模式支持从客户端到服务应用程序的类似于事件的行为。

发送一条消息而未接收任何消息的消息交换无法支持指定非 void 的返回值的服务操作;在这种情况下,将引发一个 InvalidOperationException 异常。

没有返回消息还意味着可能没有返回表明处理或通信中任何错误的 SOAP 错误。(操作是单向操作而要求双工消息交换模式时,会产生通信错误信息。)

若要为返回 void 的操作指定单向消息交换,请将 IsOneWay 属性设置为 true,如下面的 C# 代码示例所示。

[OperationContractAttribute(IsOneWay = true)] void Hello(string greeting);

此方法与前面的请求－答复示例相同,但是将 IsOneWay 属性设置为 true 意味着尽管方法相同,服务操作也不会发送返回消息,而客户端将在出站消息抵达通道层时立即返回。

8. 双工

双工模式的特点是,无论使用单向消息发送还是请求－答复消息发送方式,服务和客户端均能够独立地向对方发送消息。对于必须直接与客户端通信或向消息交换的任意一方提供异步体验(包括类似于事件的行为)的服务来说,这种双向通信形式非常有用。

由于存在与客户端通信的附加机制,双向模式比请求－答复或单向模式要略为复杂。

若要设计双工协定,还必须设计回调协定,并将该回调协定的类型分配给标记服务协定的 CallbackContract 属性(Attribute)的 ServiceContractAttribute 属性(Property)。

若要实现双工模式,必须创建第二个接口,该接口包含在客户端调用的方法声明。

小心

当服务接收双工消息时,它会在该传入消息中查找 ReplyTo 元素,以确定要发送答复的位置。如果用于接收消息的通道不安全,则不受信任的客户端可能使用目标计算机的 ReplyTo 发送恶意消息,从而导致该目标计算机发生拒绝服务(DOS)。

9. Out 和 Ref 参数

大部分情况下,用户可以使用 in 参数(ByVal 中为 Visual Basic)、out 和 ref 参数(ByRef 中为 Visual Basic)。由于 out 和 ref 参数都指示数据从操作返回的,类似如下的操作签名会指定需要请求－答复操作,即使操作签名返回 void 也是如此。

[ServiceContractAttribute] public interface IMyContract
{[OperationContractAttribute] public void PopulateData (ref CustomDataType data); }

唯一的例外是当用户的签名具有特定结构时。例如只有当用于声明操作的方法返回 NetMsmqBinding 时,才能使用 void 绑定与客户端通信;不能有任何输出值,无论它是返回值、ref,还是 out 参数。

此外,使用 out 或 ref 参数要求操作具有基础响应消息,才可以将已修改的对象传回。如果操作是单向操作,则将在运行时引发 InvalidOperationException 异常。

10. 指定协定上的消息保护级别

设计协定时,还必须确定实现用户的协定的服务的消息保护级别。仅当消息安全应用于协定终结点中的绑定时,才有必要这么做。如果绑定将安全关闭(也就是说,如果系统提供的绑定将 System. ServiceModel. SecurityMode 设置为值 SecurityMode. None),则不必确定协定的消息保护级别。大部分情况下,系统提供的绑定应用的是消息级别的安全,可提供充分的保护级别,不必考虑每个操作或每条消息的保护级别。

保护级别是一个值,它指定了支持服务的消息(或消息部分)是进行签名、签名并加密,还是未经签名或加密即发送。可以在以下多个范围设置保护级别:服务级别、针对特定操作、针对该操作内的消息或消息部分。在一个范围设置的值会成为比该范围小的范围的默认值(除非显式重写该值)。如果绑定配置无法提供协定中要求的最小保护级别,则将引发异常。如果未在协定上显式设置保护级别值,并且绑定具有消息安全,则绑定配置将控制所有消息的保护级别。这是默认行为。

重要

确定是否将协定的各种范围显式设置为小于 ProtectionLevel. EncryptAndSign 的完全保护级别的过程,通常就是牺牲部分程度的安全性来换取改善性能的过程。在这种情况下,决策必须围绕用户的操作以及它们所交换的数据值进行。

例如,下面的代码示例不会设置协定的 ProtectionLevel 或 ProtectionLevel 属性。

[ServiceContract]public interface ISampleService
{[OperationContractAttribute]public string GetString();
[OperationContractAttribute] public int GetInt();}

当使用默认的 ISampleService(其默认的 WSHttpBinding 是 System. ServiceModel. SecurityMode)在终结点中与 Message 实现交互时,将对所有消息加密并签名,因为这是默认的保护级别。但是,当 ISampleService 服务与默认 BasicHttpBinding(其默认的 SecurityMode 是 None)一起使用时,所有消息都将以文本的形式发送,这是因为此绑定没有安全性,因此将忽略保护级别(也就是说,不会对消息加密和签名)。如果 SecurityMode 更改为 Message,则将对这些消息加密并签名(因为它现在将成为绑定的默认保护级别)。

如果要显式指定或调整协定的保护要求,请将 ProtectionLevel 属性(或较小范围的任意 ProtectionLevel 属性)设置为用户的服务协定所要求的级别。在这种情况下,使用显式设置要求绑定在所用的最小范围内支持该设置。例如下面的代码示例为 ProtectionLevel 操作显式指定一个 GetGuid 值。

[ServiceContract]
public interface IExplicitProtectionLevelSampleService
{[OperationContractAttribute] public string GetString();
[OperationContractAttribute(ProtectionLevel = ProtectionLevel.None)]public int GetInt ( ); [OperationContractAttribute (ProtectionLevel = ProtectionLevel. EncryptAndSign)] public int GetGuid();}

实现此 IExplicitProtectionLevelSampleService 协定并且具有使用默认 WSHttpBinding(其默认的 System. ServiceModel. SecurityMode 是 Message)的终结点的服务具有以下行为:

(1) 对 GetString 操作消息加密并签名；
(2) GetInt 操作消息以未加密且未签名文本(即纯文本)的形式发送；
(3) GetGuid 操作 System.Guid 将在一条已加密且签名的消息中返回。

11. 其他操作签名需求

某些应用程序功能要求特定种类的操作签名。例如 NetMsmqBinding 绑定支持持久性服务和客户端，即应用程序可以在通信期间重新启动，并在其停止的位置处拾取，不会遗漏任何消息（有关详细信息，请参见 WCF 中的队列）。但是持久性操作只能接受一个 in 参数，并且没有返回值。

另一个示例是在操作中使用 Stream 类型。由于 Stream 参数包括整个消息正文，如果输入或输出(也就是 ref 参数、out 参数或返回值)的类型为 Stream，则它必须是在操作中指定的唯一输入或输出。此外，参数或返回类型必须是 Stream、System.ServiceModel.Channels.Message 或 System.Xml.Serialization.IXmlSerializable。

12. 名称、命名空间和混淆处理

在将协定转换为 WSDL 以及创建和发送协定消息时，协定与操作的定义中的 .NET 类型的名称和命名空间意义重大，因此强烈建议使用所有支持协定属性（Attribute）（如 Name、Namespace、ServiceContractAttribute、OperationContractAttribute 和其他协定属性（Attribute））的 DataContractAttribute 和 DataMemberAttribute 属性（Property）显式设置服务协定名称和命名空间。

这样做的一个原因在于，如果未显式设置名称和命名空间，则在程序集上使用 IL 混淆处理时会改变协定类型名称和命名空间，并产生已修改的 WSDL 以及通常会失败的网络交换。如果未显式设置协定名称和命名空间，但确实想使用模糊处理，请使用 ObfuscationAttribute 和 ObfuscateAssemblyAttribute 属性，以防止修改协定类型名称和命名空间。

1.3.3 实现服务协定

服务是一个类，会在一个或多个终结点公开客户端可用的功能。编写一个实现 Windows Communication Foundation (WCF) 协定的类可创建服务。可以通过两种方法执行此操作：将协定单独定义为接口，然后创建一个实现该接口的类；通过将 ServiceContractAttribute 属性放在该类本身，将 OperationContractAttribute 属性放在服务的客户端可用的方法上，来直接创建类和协定。

1. 创建服务类

下面的示例是实现已单独定义的 IMath 协定的服务。
```
[ServiceContract]
```

```
public interface IMath
{[OperationContract]    double Add(double A, double B);
   [OperationContract]  double Multiply (double A, double B); }
public class MathService : IMath
{  public double Add (double A, double B) { return A + B; }
   public double Multiply (double A, double B) { return A * B; } }
```

或者服务可以直接公开协定。下面的示例是定义和实现 MathService 协定的服务类：

```
[ServiceContract]class MathService
{[OperationContract] public double Add(double A, double B) { return A + B; }
 [OperationContract]   private double Multiply (double A, double B) { return A * B; } }
```

请注意，上面的服务公开不同的协定，因为协定名称是不同的。在第一个示例中，公开的协定命名为"IMath"，而在第二个示例中，协定命名为"MathService"。

您可以在服务和操作实现级别设置一些配置，如并发性和实例化。

在实现服务协定后，必须为该服务创建一个或多个终结点。

1.4 服务配置

在设计和实现服务协定后，即可配置服务。在其中可以定义和自定义如何向客户端公开服务，包括指定可以找到服务的地址、服务用于发送和接收消息的传输和消息编码，以及服务需要的安全类型。

此处使用的配置包括在代码中以强制方式或通过使用配置文件定义和自定义服务的各个方面的所有方法，例如指定其终结点地址、使用的传输及其安全方案。实际上，编写配置是 WCF 应用程序编程的主要部分。

1.4.1 使用配置文件配置服务

通过使用配置文件配置 Windows Communication Foundation（WCF）服务，可在部署时而非设计时提供终结点和服务行为数据的灵活性。本部分概述了当前可用的主要技术。

可使用 WCF 配置技术对 .NET Framework 服务进行配置。通常情况下，向承载 WCF 服务的 Internet 信息服务（IIS）网站的 Web.config 文件添加 XML 元素。通过这些元素，可以逐台计算机更改详细信息，例如终结点地址（用于与服务进行通信的实际地址）。此外，WCF 包括几个系统提供的元素，可用于快速选择服务的最基本的功能。从 .NET Framework 版本 4 开始，WCF 附带一个新的默认配置模型，该模型简化了 WCF 配置要求。如果没有为特定服务提供任何 WCF 配置,运行时将自动使用一些标准终结点和默认绑定/行为配置服务。实际上,编写配置是 WCF 应用程序编程的主要部分。

重要

在部署并行方案(其中部署了服务的两个不同版本)时,必须指定配置文件中引用的程序集的部分名称。这是因为配置文件将在服务的所有版本间共享,并可在不同版本的 .NET Framework 下运行。

1. System. Configuration:Web. config 和 App. config

WCF 使用 .NET Framework 的 System. Configuration 配置系统。

在 Visual Studio 中配置服务时,使用 Web. config 文件或 App. config 文件指定设置。配置文件名称的选择由为服务选择的宿主环境确定。如果正在使用 IIS 来承载服务,则使用 Web. config 文件;如果正在使用任何其他宿主环境,则使用 App. config 文件。

在 Visual Studio 中,名为 App. config 的文件可用于创建最终的配置文件。实际用于配置的最终名称取决于程序集名称。例如名为"Cohowinery. exe"的程序集具有的最终配置文件名称为 Cohowinery. exe. config。但是,只需要修改 App. config 文件。在编译时,对该文件所做的更改会自动应用于最终应用程序配置文件。

在使用 App. config 文件的过程中,当应用程序启动并应用配置时,文件配置系统会将 App. config 文件与 Machine. config 文件的内容合并。此机制允许在 Machine. config 文件中定义计算机范围的设置。可以使用 App. config 文件重写 Machine. config 文件的设置;也可以锁定 Machine. config 文件中的设置以应用它们。对于 Web. config,配置系统会将应用程序目录之下的所有目录中的 Web. config 文件合并到要应用的配置中。

2. 配置文件的主要部分

配置文件中的主要部分包括以下元素。

```
< system.ServiceModel >
    < services >
        < service >
            < endpoint />
        < /service >
    < /services >
    < bindings >
        < binding >
            < /binding >
    < /bindings >
    < behaviors >
        < behavior >
```

```
        </behavior>
    </behaviors>
</system.ServiceModel>
```

备注

绑定部分和行为部分是可选的,只在需要时才包括。

(1) <服务>元素

服务元素包含应用程序承载的所有服务的规范。从.NET Framework 4 中的简化配置模型开始,此部分是可选的。

每个服务都具有以下属性:

◇ name。指定提供服务协定的实现的类型。这是完全限定名称,其中包含命名空间、句点和类型名称。例如"MyNameSpace.myServiceType"。

◇ behaviorConfiguration。指定一个在 behavior 元素中找到的 behaviors 元素的名称。指定的行为控制操作,例如服务是否允许模拟。如果它的值是空的,或者未提供任何 behaviorConfiguration,则向服务中添加默认服务行为集。

(2) <终结点>元素

每个终结点都需要以下属性表示的地址、绑定和协定:

◇ address。指定服务的统一资源标识符(URI),它可以是一个绝对地址,或是一个相对于服务基址给定的地址。如果设置为空字符串,则指示在创建服务的 ServiceHost 时,终结点在指定的基址上可用。

◇ binding。通常,指定一个类似 WSHttpBinding 的系统提供的绑定,但也可以指定一个用户定义的绑定。指定的绑定确定传输协议类型、安全和使用的编码,以及是否支持或启用可靠会话、事务或流。

◇ bindingConfiguration。如果必须修改绑定的默认值,则可通过在 binding 元素中配置相应的 bindings 元素来执行此操作。此属性应赋予与用于更改默认值的 name 元素的 binding 属性相同的值。如果未提供任何名称,或者在绑定中未指定任何 bindingConfiguration,则在终结点中使用绑定类型的默认绑定。

◇ contract。指定定义协定的接口。这是在由 name 元素的 service 属性指定的公共语言运行库(CLR)类型中实现的接口。

◇ <终结点>元素引用

<client>的<endpoint>指定通道终结点的协定、绑定和地址属性,客户端使用通道终结点与服务器上的服务终结点连接,其引用语法如下:

```
<endpoint address = "String"
    behaviorConfiguration = "String"
    binding = "String"
```

```
            bindingConfiguration = "String"
            contract = "String"   endpointConfiguration = "String"   kind = "String"
            name = "String"
</endpoint>
```

(3) <绑定>元素

bindings 元素包含可由任何服务中定义的任何终结点使用的所有绑定的规范。

在 binding 元素中包含的 bindings 元素可以是系统提供的绑定之一,也可以是自定义绑定。binding 元素具有 name 属性,此属性将绑定与 bindingConfiguration 元素的 endpoint 属性中指定的终结点相关联。如果未指定任何名称,则该绑定对应于该绑定类型的默认值。

可以使用 binding 元素来配置 Windows Communication Foundation (WCF) 提供的不同类型的预定义绑定。

(4) 系统提供的绑定

系统提供的绑定可以消除 WCF 消息堆栈的复杂性。使用系统提供的绑定的应用程序不需要对堆栈的完全控制权限。在每个系统提供的绑定上公开的属性最适合绑定所针对的使用方案。

每个系统提供的绑定的配置节可以定义用于配置此绑定的一些配置。每个配置均由唯一的名称进行标识。

无法向系统提供的绑定添加元素或属性。为此,应该按照本主题的"自定义绑定"节中的描述来实现自定义绑定。可以定义一个自定义绑定,该自定义绑定将完全模仿系统提供的绑定,并添加用户应用程序希望获得其控制权限的一些设置。

(5) 自定义绑定

自定义绑定提供了对 WCF 消息堆栈的完全控制。单个绑定按照堆栈元素在堆栈上出现的顺序来指定它们的配置元素,从而定义消息堆栈。每个元素都定义和配置了该堆栈的一个元素。在每个自定义绑定中,必须有且只能有一个 transport 元素。如果没有该元素,消息堆栈将是不完整的。

元素在堆栈中出现的顺序非常重要,因为在将操作应用于消息时会采用该顺序。建议的堆栈元素顺序如下:

①事务(可选);
②可靠消息(可选);
③安全(可选);
④编码器;
⑤传输。

自定义绑定由其 name 特性来标识。

(6) <行为>元素

这是定义服务行为的 behavior 元素的容器元素。

此元素定义名为 endpointBehaviors 和 serviceBehaviors 的两个子集合。每个集合分别定义终结点和服务所使用的行为元素。每个行为元素由其唯一的 name 属性标识。从 .NET Framework 4 开始,不要求绑定和行为具有名称。行为元素的语法如下:

```
<behaviors>
    <serviceBehaviors>
    </serviceBehaviors>
    <endpointBehaviors>
    </endpointBehaviors>
</behaviors>
```

(7) 如何使用绑定和行为配置

在 WCF 中,通过在配置中使用引用系统,可以很方便地在终结点之间共享配置。与绑定相关的配置值在 bindingConfiguration 部分的 <binding> 元素中进行分组,而不是直接将配置值分配到终结点。绑定配置是一组命名的绑定设置。然后,终结点可以通过名称来引用 bindingConfiguration。绑定和行为配置的具体语法如下:

```
<?xml version = "1.0" encoding = "utf-8"?>
<configuration>
<system.serviceModel>
  <bindings>
    <basicHttpBinding>
      <binding name = "myBindingConfiguration1" closeTimeout = "00:01:00" />
      <binding name = "myBindingConfiguration2" closeTimeout = "00:02:00" />
      <binding closeTimeout = "00:03:00" />
    <!-- Default binding for basicHttpBinding -->
    </basicHttpBinding>
  </bindings>
  <services>
    <service name = "MyNamespace.myServiceType">
    <endpoint
       address = "myAddress" binding = "basicHttpBinding"
       bindingConfiguration = "myBindingConfiguration1"
       contract = "MyContract"  />
    <endpoint
       address = "myAddress2" binding = "basicHttpBinding"
       bindingConfiguration = "myBindingConfiguration2"
```

```
            contract = "MyContract" />
        <endpoint
            address = "myAddress3" binding = "basicHttpBinding"
            contract = "MyContract" />
        </service>
      </services>
    </system.serviceModel>
</configuration>
```

在 name 元素中设置 bindingConfiguration 的 <binding>。name 必须是绑定类型的范围内的唯一字符串——在这种情况下 < basicHttpBinding >，或者是引用默认绑定的空值。通过将 bindingConfiguration 属性设置为此字符串，终结点链接到该配置。

以相同方式实现 behaviorConfiguration，如以下示例中所示。

```
<?xml version = "1.0" encoding = "utf-8"?>
<configuration>
  <system.serviceModel>
    <behaviors>
      <endpointBehaviors>
        <behavior name = "myBehavior">
          <callbackDebug includeExceptionDetailInFaults = "true" />
        </behavior>
      </endpointBehaviors>
      <serviceBehaviors>
        <behavior>
          <serviceMetadata httpGetEnabled = "true" />
        </behavior>
      </serviceBehaviors>
    </behaviors>
    <services>
      <service name = "NewServiceType">
        <endpoint
            address = "myAddress3" behaviorConfiguration = "myBehavior"
            binding = "basicHttpBinding"
            contract = "MyContract" />
        </service>
      </services>
    </system.serviceModel>
</configuration>
```

请注意,向服务中添加了默认服务行为集。此系统允许终结点共享公共配置而不用重新定义设置。如果需要界定范围,则在 Machine.config 中创建绑定或行为配置。配置设置在所有 App.config 文件中可用。通过 Configuration Editor Tool（SvcConfigEditor.exe）可以很方便地创建配置。

(9) 行为合并

当需要统一使用一组公共行为时,利用行为合并功能,可更加轻松地管理行为。此功能允许在配置层次结构的各个层上指定行为,并使服务能够从配置层次结构的多个层继承行为。为了演示此功能的工作方式,假定 IIS 中包含以下虚拟目录布局:

~\Web.config ~\Service.svc ~\Child\Web.config ~\Child\Service.svc

~\Web.config 文件包含以下内容：

```
<configuration>
  <system.serviceModel>
    <behaviors>
      <serviceBehaviors>
        <behavior>
          <serviceDebug includeExceptionDetailInFaults = "True" />
        </behavior>
      </serviceBehaviors>
    </behaviors>
  </system.serviceModel>
</configuration>
```

具有一个位于 ~\Child\Web.config 并包含以下内容的子 Web.config：

```
<configuration>
  <system.serviceModel>
    <behaviors>
      <serviceBehaviors>
        <behavior>
          <serviceMetadata httpGetEnabled = "True" />
        </behavior>
      </serviceBehaviors>
    </behaviors>
  </system.serviceModel>
</configuration>
```

位于 ~\Child\Service.svc 的服务的行为如同该服务具有 serviceDebug 和 serviceMetadata 行为一样。位于 ~\Service.svc 的服务只具有 serviceDebug 行为。发生的情况是,名称相同的两个行为集合(此情况下为空字符串)将合并。

可以通过使用集合<清除>标记来定义清除行为,也可从集合中移除各个行为<删除>标记。例如子服务中的以下两个配置结果只具有 serviceMetadata 行为:

```
< configuration >
  < system.serviceModel >
    < behaviors >
      < serviceBehaviors >
        < behavior >
          < remove name = "serviceDebug" />
          < serviceMetadata httpGetEnabled = "True" />
        < /behavior >
      < /serviceBehaviors >
    < /behaviors >
  < /system.serviceModel >
< /configuration >
< configuration >
  < system.serviceModel >
    < behaviors >
      < serviceBehaviors >
        < behavior >
          < clear />
          < serviceMetadata httpGetEnabled = "True" />
        < /behavior >
      < /serviceBehaviors >
    < /behaviors >
  < /system.serviceModel >
< /configuration >
```

对无名称行为集合执行行为合并(如上所示),并对命名行为集合执行行为合并。行为合并在 IIS 宿主环境中进行,在该环境中,Web.config 文件将以分层方式与根 Web.config 文件和 machine.config 合并。但行为合并也可在应用程序环境中进行,在该环境中,machine.config 可与 App.config 文件合并。

行为合并适用于配置中的终结点行为和服务行为。

如果子行为集合包含一个已显示在父行为集合中的行为,则子行为将重写父行为,因此如果父行为集合具有 < serviceMetadata httpGetEnabled = "False" />,而子行为集合具有 < serviceMetadata httpGetEnabled = "True" />,则子行为将重写行为集合中的父行为并且 httpGetEnabled 将为"true"。

1.4.2 代码中配置 WCF 服务

通过 WCF,开发人员可以使用配置文件或代码来配置服务。当部署之后需要对服务进行配置时,配置文件十分有用。在使用配置文件时,IT 专业人员只需要更新配置文件,无须重新编译。不过,配置文件可能十分复杂,难以维护。不支持对配置文件进行调试,并且将按名称来引用配置元素,这使得配置文件的创作较困难且易于出错。通过 WCF,还可以使用代码来配置服务。在早期版本的 WCF(4.0 及更早版本)中,用代码来配置服务在自承载方案中十分方便,可以在调用 ServiceHost.Open 之前,使用 ServiceHost 类配置终结点和行为。但是,在 Web 承载方案中,不具备针对 ServiceHost 类的直接访问权限。若要配置 Web 承载的服务,需要创建 System.ServiceModel.ServiceHostFactory,后者会创建 ServiceHostFactory 并执行任何所需的配置。从 .NET 4.5 起,WCF 提供了一种使用代码来配置自承载服务和 Web 承载服务的更方便的方法。

Configure 方法

只需在服务实现类中使用以下签名定义名为 Configure 的公共静态方法:

```
public static void Configure(ServiceConfiguration config)
```

Configure 方法采用 ServiceConfiguration 实例,使开发者可以添加终结点和行为。在打开服务主机之前,由 WCF 调用此方法。定义后,将忽略 app.config 或 web.config 文件中指定的任何服务配置设置。

下面的代码段阐释如何定义 Configure 方法和添加服务终结点、终结点行为和服务行为:

```
public class Service1 : IService1
    {
        public static void Configure(ServiceConfiguration config)
        {
            ServiceEndpoint se = new ServiceEndpoint(new ContractDescription("IService1"), new BasicHttpBinding(), new EndpointAddress("basic"));
            se.Behaviors.Add(new MyEndpointBehavior());
            config.AddServiceEndpoint(se);
            config.Description.Behaviors.Add(new ServiceMetadataBehavior{ HttpGetEnabled = true });
            config.Description.Behaviors.Add(new ServiceDebugBehavior{ IncludeExceptionDetailInFaults = true });
        }
        public string GetData(int value)
        {
```

```csharp
            return string.Format("You entered: {0}", value);
        }
        public CompositeType GetDataUsingDataContract(CompositeType composite)
        {
            if (composite == null)
            {
                throw new ArgumentNullException("composite");
            }
            if (composite.BoolValue)
            {
                composite.StringValue += "Suffix";
            }
            return composite;
        }
    }
```

要为服务启用协议(如 https),可以显式添加使用协议的终结点,也可以通过调用 ServiceConfiguration.EnableProtocol(Binding)自动添加终结点,这样可为与协议兼容的每个基址和定义的每个服务协定添加终结点。下面的代码演示如何使用 ServiceConfiguration. EnableProtocol 方法:

```csharp
public class Service1 : IService1
{
    public string GetData(int value);
    public static void Configure(ServiceConfiguration config)
    {
        //Enable "Add Service Reference" support
        config.Description.Behaviors.Add(new ServiceMetadataBehavior{ HttpGetEnabled = true });
        //set up support for http, https, net.tcp, net.pipe
        config.EnableProtocol(new BasicHttpBinding());
        config.EnableProtocol(new BasicHttpBinding());
        config.EnableProtocol(new NetTcpBinding());
        config.EnableProtocol(new NetNamedPipeBinding());
        //add an extra BasicHttpBinding endpoint at http:///basic
        config.AddServiceEndpoint(typeof(IService1), new BasicHttpBinding(),"basic");
    }
}
```

可以从默认应用程序的配置文件来调用需要加载的服务配置 LoadFromConfiguration，然后更改设置。LoadFromConfiguration()类还允许从集中式配置加载配置。下面的代码演示如何实现这一点：

```
public class Service1 : IService1
{
    public void DoWork();
    public static void Configure(ServiceConfiguration config)
    {
        config.LoadFromConfiguration(ConfigurationManager.OpenMappedExeConfiguration
(new ExeConfigurationFileMap | ExeConfigFilename = @ "c:\sharedConfig\MyConfig.config" |, ConfigurationUserLevel.None));
    }
}
```

重要

请注意，LoadFromConfiguration 忽略 < host > 中的设置 < service > 标记 < system.serviceModel >。从概念上讲，< host > 即将主机配置成无服务配置，它获取并加载之前执行此配置方法。

1.5 客户端生成

1.5.1 WCF 客户端概述

本节描述客户端应用程序可以做什么，如何配置、创建和使用 WCF 客户端，以及如何保护客户端应用程序。

1. 使用 WCF 客户端对象

客户端应用程序是使用 WCF 客户端与其他应用程序通信的托管应用程序。若要为 WCF 服务创建一个客户端应用程序，则需要执行下列步骤：

(1) 获取服务终结点的服务协定、绑定以及地址信息。
(2) 使用该信息创建 WCF 客户端。
(3) 调用操作。
(4) 关闭该 WCF 客户端对象。

以下部分将讨论上述这些步骤，并简单介绍以下问题：

◇ 处理错误。

◇ 配置和保护客户端。
◇ 为双工服务创建回调对象。
◇ 异步调用服务。
◇ 使用客户端通道调用服务。

获取服务协定、绑定和地址

在 WCF 中,服务和客户端使用托管属性、接口和方法对协定进行建模。若要连接客户端应用程序中的服务,则需要获取该服务协定的类型信息。通常情况下,使用 ServiceModel 元数据实用工具(Svcutil.exe)可以将下载的元数据转换为托管代码文件中所选择的语言,并创建一个客户端用于配置 WCF 客户端对象的应用程序配置文件。例如如果准备创建一个 WCF 客户端对象来调用 MyCalculatorService,并且知道该服务的元数据已在 http://computerName/MyCalculatorService/Service.svc? wsdl 中发布,则下面的代码示例演示如何使用 Svcutil.exe 获取一个包含托管代码中的服务协定的 ClientCode.vb 文件。

svcutil /language: vb /out: ClientCode.vb /config: app.config http://computerName/MyCalculatorService/Service.svc? wsdl

可以将此协定代码编译为客户端应用程序或另一个程序集,然后,客户端应用程序可以使用该程序集创建一个 WCF 客户端对象。可以使用配置文件配置客户端对象以与服务正确连接。

2. 创建一个 WCF 客户端对象

WCF 客户端是表示某个 WCF 服务的一个本地对象,客户端可以使用这种表示形式与远程服务进行通信。WCF 客户端类型可实现目标服务协定,因此在创建一个服务协定并配置它之后,就可以直接使用该客户端对象调用服务操作。WCF 运行时将方法调用转换为消息,将它们发送到服务,侦听回复,并将这些值作为返回值返回给 WCF 客户端对象或 out 或 ref 参数。

还可以使用 WCF 客户端通道对象连接和使用服务。

创建一个新的 WCF 对象:

为了演示如何使用 ClientBase < TChannel > 类,现假设已从服务应用程序生成了下面的简单服务协定。

备注

如果使用 Visual Studio 创建 WCF 客户端,对象将自动加载到对象浏览器并添加到项目的服务引用时。

```
[System.ServiceModel.ServiceContractAttribute(
  Namespace = "http://microsoft.wcf.documentation")]
public interface ISampleService
```

```
    [System.ServiceModel.OperationContractAttribute(
     Action = "http://microsoft.wcf.documentation/ISampleService/SampleMethod",
     ReplyAction = "http://microsoft.wcf.documentation/ISampleService/SampleMethodResponse")]
    [System.ServiceModel.FaultContractAttribute(
     typeof(microsoft.wcf.documentation.SampleFault),
      Action = " http://microsoft.wcf.documentation/ISampleService/SampleMethodSampleFaultFault")]
    string SampleMethod(string msg);}
```

如果不是使用 Visual Studio,则请检查已生成的协定代码以查找扩展 ClientBase<TChannel> 的类型以及服务协定接口 ISampleService。在这种情况下,该类型看上去类似下列代码:

```
[System.CodeDom.Compiler.GeneratedCodeAttribute("System.ServiceModel", "3.0.0.0")]
public partial class SampleServiceClient : System.ServiceModel.ClientBase<ISampleService>, ISampleService
{    public SampleServiceClient()
 {   }
    public SampleServiceClient(string endpointConfigurationName)
     :
        base(endpointConfigurationName)
 {   }
     public SampleServiceClient ( string endpointConfigurationName, string remoteAddress) : base(endpointConfigurationName, remoteAddress)
 {   }
     public SampleServiceClient ( string endpointConfigurationName, System.ServiceModel.EndpointAddress remoteAddress)       :
        base(endpointConfigurationName, remoteAddress)
 {   }
    public SampleServiceClient(System.ServiceModel.Channels.Binding binding, System.ServiceModel.EndpointAddress remoteAddress)       :
        base(binding, remoteAddress)
 {   }
    public string SampleMethod(string msg)
    {
        return base.Channel.SampleMethod(msg);
    }}
```

可以通过使用其中一个构造函数将此类创建为一个本地对象,并对该本地对象进行配置,然后使用它连接到 ISampleService 类型的服务。

建议首先创建 WCF 客户端对象,然后使用该对象并在一个单独的 try/catch 块内将它关闭。不应该使用 using 语句(Using 中的 Visual Basic),因为该语句可以屏蔽处于某些失败模式的异常。

3. 协定、绑定和地址

在可以创建 WCF 客户端对象之前,必须配置客户端对象。具体而言,它必须有一个服务终结点使用。终结点由服务协定、绑定和地址组成。通常情况下,此信息是否位于<终结点>客户端应用程序配置文件,例如 Svcutil.exe 工具生成,并创建客户端自动加载中的元素对象。两种 WCF 客户端类型都具有能够以编程方式指定此信息的重载。

例如,上述示例中所使用的 ISampleService 的生成的配置文件包含以下终结点信息。

```
<configuration>
    <system.serviceModel>
        <bindings>
            <wsHttpBinding>
<binding name = "WSHttpBinding_ISampleService" closeTimeout = "00:01:00"
openTimeout = "00:01:00" receiveTimeout = "00:01:00" sendTimeout = "00:01:00"
bypassProxyOnLocal = "false" transactionFlow = "false" hostNameComparisonMode
= "StrongWildcard" maxBufferPoolSize = "524288" maxReceivedMessageSize = "65536"
messageEncoding = "Text" textEncoding = "utf - 8" useDefaultWebProxy = "true"
allowCookies = "false">
    <readerQuotas maxDepth = "32" maxStringContentLength = "8192" maxArrayLength
= "16384" maxBytesPerRead = "4096" maxNameTableCharCount = "16384" />
    <reliableSession ordered = "true" inactivityTimeout = "00:10:00" enabled = "
false" />
    <security mode = "Message"> <transport clientCredentialType = "None"
proxyCredentialType = "None" realm = "" /> <message clientCredentialType = "
Windows" negotiateServiceCredential = "true"
algorithmSuite = "Default" establishSecurityContext = "true" /> </security>
        </binding>
            </wsHttpBinding> </bindings> <client>
                <endpoint address = "http://localhost:8080/SampleService" binding
= "wsHttpBinding" bindingConfiguration = "WSHttpBinding_ISampleService" contract
= "ISampleService" name = "WSHttpBinding_ISampleService"> </endpoint> </client
> </system.serviceModel> </configuration>
```

此配置文件在 <client> 元素中指定目标终结点。

4. 调用操作

创建并配置了客户端对象后,请创建一个 try/catch 块,如果该对象是本地对象,则以相同的方式调用操作,然后关闭 WCF 客户端对象。当客户端应用程序调用第一个操作时,WCF 将自动打开基础通道,并在回收对象时关闭基础通道。(或者,还可以在调用其他操作之前或之后显式打开和关闭该通道)

例如如果具有以下服务协定:

```
namespace Microsoft.ServiceModel.Samples
{using System; using System.ServiceModel;
    [ServiceContract(Namespace = "http://Microsoft.ServiceModel.Samples")]
    public interface ICalculator
    {[OperationContract]  double Add(double n1, double n2);
     [OperationContract]  double Subtract(double n1, double n2);
     [OperationContract]  double Multiply(double n1, double n2);
     [OperationContract]  double Divide(double n1, double n2);
    }}
```

可以通过创建一个 WCF 客户端对象并调用其方法来调用操作,如以下代码示例所示。请注意,打开、调用和关闭 WCF 客户端对象发生在一个 try/catch 块内。

```
CalculatorClient wcfClient = new CalculatorClient();
try{   Console.WriteLine(wcfClient.Add(4, 6));
    wcfClient.Close();}
catch (TimeoutException timeout)
{     wcfClient.Abort();}
catch (CommunicationException commException)
{     wcfClient.Abort();}
```

5. 处理错误

当打开基础客户端通道(无论是通过显式打开还是通过调用操作自动打开)、使用客户端或通道对象调用操作,或关闭基础客户端通道时,都会在客户端应用程序中出现异常。除了由操作返回的 SOAP 错误导致引发的任何 System.TimeoutException 对象外,建议至少将应用程序设置为能够处理可能的 System.ServiceModel.CommunicationException 和 System.ServiceModel.FaultException 异常。操作协定中指定的 SOAP 错误将作为 System.ServiceModel.FaultException<TDetail> 在客户端应用程序中引发,此异常中的类型参数为 SOAP 错误的详细信息类型。

6. 配置和保护客户端

若要配置客户端,应首先为客户端或通道对象加载目标终结点信息,通常是从配置文件中加载该信息,但是也可以使用客户端构造函数和属性以编程方式加载。但是,若要启用特定的客户端行为或实施一些安全方案还需要执行其他配置步骤。

例如服务协定的安全要求已在服务协定接口中声明,并且如果 Svcutil.exe 已创建了一个配置文件,则该文件通常会包含一个能够支持服务安全要求的绑定。但是在某些情况中,可能需要更多的安全配置,例如配置客户端凭据。

此外,在客户端应用程序中还可以启用一些自定义修改,例如自定义运行时行为。

7. 为双工服务创建回调对象

双工服务指定一个回调协定,客户端应用程序必须实现该协定以便提供一个该服务能够根据协定要求调用的回调对象。虽然回调对象不是完整的服务(例如无法使用回调对象启动一个通道),但是为了实现和配置,这些回调对象可以被视为一种服务。

双工服务的客户端必须满足以下三个条件:

(1)实现一个回调协定类。

(2)创建回调协定实现类的一个实例,并使用该实例创建传递给 System.ServiceModel.InstanceContext 客户端构造函数的 WCF 对象。

(3)调用操作并处理操作回调。

双工 WCF 客户端对象除了会公开支持回调所必需的功能(包括回调服务的配置)以外,其他的功能和它们的非双工对应项相同。

例如可以通过使用回调类的 System.ServiceModel.CallbackBehaviorAttribute 属性(Attribute)的属性(Property),控制回调对象运行时行为的各个方面。另一个示例是使用 System.ServiceModel.Description.CallbackDebugBehavior 类将异常信息返回给调用回调对象的服务。有关详细信息,请参见双工服务。

在运行 Internet 信息服务(IIS)5.1 的 Windows XP 计算机上,双工客户端必须使用 System.ServiceModel.WSDualHttpBinding 类指定一个客户端基址,否则会引发异常。下面的代码示例演示如何在代码中执行此操作:

```
WSDualHttpBinding dualBinding = new WSDualHttpBinding();
EndpointAddress endptadr = new EndpointAddress（"http://localhost:12000/DuplexTestUsingCode/Server"）;
dualBinding.ClientBaseAddress = new Uri（"http://localhost:8000/DuplexTestUsingCode/Client/"）;
```

下面的代码演示如何在配置文件中执行此操作:

```
<client>
  <endpoint
    name="ServerEndpoint"
    address="http://localhost:12000/DuplexUsingConfig/Server"
```

```
      bindingConfiguration = "WSDualHttpBinding_IDuplex"
      binding = "wsDualHttpBinding"
      contract = "IDuplex" />
</client>
<bindings>
  <wsDualHttpBinding>
    <binding
      name = "WSDualHttpBinding_IDuplex"
      clientBaseAddress = "http://localhost:8000/myClient/" />
  </wsDualHttpBinding>
</bindings>
```

8. 异步调用服务

如何调用操作完全取决于客户端开发人员。这是因为当在托管代码中表示组成操作的消息时,这些消息可以映射到同步或异步方法中,因此如果想要生成异步调用操作的客户端,则可以使用 Svcutil.exe 通过 /async 选项生成异步客户端代码。

9. 使用 WCF 客户端通道调用服务

WCF 客户端类型扩展 ClientBase<TChannel>,而其自身派生自 System.ServiceModel.IClientChannel 接口,从而可以公开基础通道系统。可以同时使用目标服务协定和 System.ServiceModel.ChannelFactory<TChannel> 类来调用服务。

1.5.2 WCF 客户端访问服务

创建服务之后,下一步是创建 WCF 客户端代理。客户端应用程序使用 WCF 客户端代理与服务进行通信。客户端应用程序通常导入服务的元数据来生成可用来调用该服务的 WCF 客户端代码。

1. 创建 WCF 客户端的基本步骤

(1)编译服务代码。
(2)生成 WCF 客户端代理。
(3)实例化 WCF 客户端代理。

服务模型 Metadata Utility Tool(Svcutil.exe)可以手动生成 WCF 客户端代理 ServiceModel Metadata Utility Tool(Svcutil.exe)。此外,还可以在 Visual Studio 内生成 WCF 客户端代理使用添加服务引用功能。若要使用上述两种方法中的一种生成 WCF 客户端代理,必须运行该服务。如果服务是自承载服务,则必须运行主机。如果服务是在 IIS/WAS 中承载的,则无须执行任何其他操作。

2. ServiceModel 元数据实用工具

ServiceModel 元数据实用工具（Svcutil.exe）是用于从元数据生成代码的命令行工具。下面是一个基本 Svcutil.exe 命令示例：

Svcutil.exe <service's Metadata Exchange (MEX) address or HTTP GET address>

另外，还可以使用 Svcutil.exe 来处理文件系统中的 Web 服务描述语言（WSDL）和 XML 架构定义语言（XSD）文件。

Svcutil.exe <list of WSDL and XSD files on file system>

结果是一个包含 WCF 客户端代码的代码文件，客户端应用程序可以使用这些客户端代码来调用服务。

还可以使用该工具生成配置文件。

Svcutil.exe <file1 [,file2]>

如果仅提供了一个文件名，则该文件名是输出文件的名称。如果提供了两个文件名，则第一个文件是输入配置文件，其内容将与生成的配置合并，然后写出到第二个文件中。

重要

与任何未受保护的网络请求一样，未受保护的元数据请求也会带来一定的风险：如果不能确定正在与其进行通信的终结点身份属实，那么检索的信息有可能是来自于恶意服务的元数据。

3. Visual Studio 中的"添加服务引用"功能

服务运行时，右键单击项目，它将包含 WCF 客户端代理并且选择添加服务引用。在添加服务引用对话框中想要调用并单击服务的 URL 类型转换按钮。该对话框将显示在指定的地址上提供的服务列表。双击服务可看到可用协定和操作，指定生成的代码的命名空间，然后单击确定按钮。

示例

下面的代码示例演示为服务创建的一个服务协定。

```
//Define a service contract.
[ServiceContract(Namespace = "http://Microsoft.ServiceModel.Samples")]
public interface ICalculator
{ [OperationContract]    double Add(double n1, double n2); }
```

ServiceModel 元数据实用工具以及 Visual Studio 中的"添加服务引用"功能生成以下 WCF 客户端类。该类从 ClientBase<TChannel> 泛型类继承，并实现 ICalculator 接口。该工具还生成 ICalculator 接口（此处未演示）。

```
public partial class CalculatorClient : System.ServiceModel.ClientBase<
ICalculator>, ICalculator
```

```
    public CalculatorClient()        {}
  public CalculatorClient(string endpointConfigurationName) :
        base(endpointConfigurationName)        {}
   public CalculatorClient ( string endpointConfigurationName, string
remoteAddress) : base(endpointConfigurationName, remoteAddress)        {}
  public CalculatorClient(string endpointConfigurationName,
     System.ServiceModel.EndpointAddress remoteAddress) :
        base(endpointConfigurationName, remoteAddress) {}
  public CalculatorClient(System.ServiceModel.Channels.Binding binding,
     System.ServiceModel.EndpointAddress remoteAddress) :
        base(binding, remoteAddress) {}
  public double Add(double n1, double n2)
  {      return base.Channel.Add(n1, n2); }}
```

4. 使用 WCF 客户端

若要使用 WCF 客户端,应创建 WCF 客户端的一个实例,然后调用其方法,如下面的代码所示。

```
CalculatorClient calcClient = new CalculatorClient("CalculatorEndpoint"));
double value1 = 100.00D;
double value2 = 15.99D;double result = calcClient.Add(value1, value2);
Console.WriteLine("Add({0},{1}) = {2}", value1, value2, result);
```

5. 调试客户端引发的异常

WCF 客户端所引发的许多异常是由服务上的异常所导致的。以下是这种情况的一些示例。

SocketException:现有连接被远程主机强行关闭。

CommunicationException:基础连接意外关闭。

CommunicationObjectAbortedException:套接字连接已中止。这可能是由于处理消息时出错或远程主机接收超时或者潜在的网络资源问题导致的。

当发生这些类型的异常时,解决问题的最佳方式是在服务端启用跟踪并确定服务端发生了何种异常。

1.5.4 指定客户端运行时行为

与 WCF 服务相似,可以对 WCF 客户端进行配置以修改运行时行为,以便适合客户端应用程序的需要。有三个属性可用于指定客户端运行时行为。双工客户端回调对象可以使用 CallbackBehaviorAttribute 和 CallbackDebugBehavor 属性修改其运行时行为。另一个属性

ClientViaBehavior 可用于将逻辑目标与直接网络目标分开。此外，双工客户端回调类型可以使用某些服务端行为。

1. 使用 CallbackBehaviorAttribute

可以使用 CallbackBehaviorAttribute 类来配置或扩展客户端应用程序中回调协定实现的执行行为。此属性为回调类，和 ServiceBehaviorAttribute 类执行相似的功能，不同之处在于实例化行为和事务设置。

必须将 CallbackBehaviorAttribute 类应用于实现回调协定的类。如果将其应用于非双工协定实现，则会在运行时引发 InvalidOperationException 异常。下面的代码示例演示回调对象上的一个 CallbackBehaviorAttribute 类，该类使用 SynchronizationContext 对象确定要封送到的线程，使用 ValidateMustUnderstand 属性强制执行消息验证，并使用 IncludeExceptionDetailInFaults 属性将异常作为 FaultException 对象返回给服务以便进行调试。

```
using System;using System.ServiceModel;using System.ServiceModel.Channels;
using System.Threading;
namespace Microsoft.WCF.Documentation
{
  [CallbackBehaviorAttribute(
   IncludeExceptionDetailInFaults = true,
   UseSynchronizationContext = true,
   ValidateMustUnderstand = true
  )]
  public class Client : SampleDuplexHelloCallback
  { AutoResetEvent waitHandle;
    public Client()
    {     waitHandle = new AutoResetEvent(false);    }
    public void Run()
    { SampleDuplexHelloClient wcfClient
         = new SampleDuplexHelloClient ( new InstanceContext ( this ), "WSDualHttpBinding_SampleDuplexHello");
      try
      { Console.ForegroundColor = ConsoleColor.White;
        Console.WriteLine("Enter a greeting to send and press ENTER: ");
        Console.Write("> > > ");
        Console.ForegroundColor = ConsoleColor.Green;
        string greeting = Console.ReadLine();
        Console.ForegroundColor = ConsoleColor.White;
```

```
            Console.WriteLine("Called service with: \r\n\t" + greeting);
            wcfClient.Hello(greeting);
             Console.WriteLine("Execution passes service call and moves to the
WaitHandle.");
            this.waitHandle.WaitOne();
            Console.ForegroundColor = ConsoleColor.Blue;
            Console.WriteLine("Set was called.");
            Console.Write("Press ");
            Console.ForegroundColor = ConsoleColor.Red;
            Console.Write("ENTER");
            Console.ForegroundColor = ConsoleColor.Blue;
            Console.Write(" to exit...");
            Console.ReadLine();          }
        catch (TimeoutException timeProblem)
        {        Console.WriteLine("The service operation timed out. " +
timeProblem.Message);
            Console.ReadLine();          }
        catch (CommunicationException commProblem)
    {Console.WriteLine("There was a communication problem. " + commProblem.
Message);  Console.ReadLine();    }      }
     public static void Main()
     {     Client client = new Client();    client.Run();    }
     public void Reply(string response)
     { Console.WriteLine("Received output.");
        Console.WriteLine("\r\n\t" + response);
        this.waitHandle.Set();
     }  }  }
```

2. 使用 CallbackDebugBehavior 启用托管异常信息流

可以在客户端回调对象中启用托管异常信息流,使异常信息流回到服务以便进行调试,方法是以编程方式或从应用程序配置文件中将 IncludeExceptionDetailInFaults 属性设置为 true。

将托管异常信息返回到服务可能存在安全风险,因为异常详细信息会公开与内部客户端实现有关的信息,而未经授权的服务可能会使用这些信息。此外,虽然 CallbackDebugBehavior 属性也可以通过编程方式进行设置,但在部署时容易忘记禁用 IncludeExceptionDetailInFaults。

由于涉及一些安全问题,因此强烈建议用户:

(1)使用应用程序配置文件将 IncludeExceptionDetailInFaults 属性的值设置为 true。
(2)仅在受控调试方案中才能这样做。

下面的代码示例演示一个客户端配置文件,该文件指示 WCF 从 SOAP 消息中的客户端回调对象返回托管异常信息。

```xml
<client>
    <endpoint
      address = "http://localhost:8080/DuplexHello"
      binding = "wsDualHttpBinding"
      bindingConfiguration = "WSDualHttpBinding_SampleDuplexHello"
      contract = "SampleDuplexHello"
      name = "WSDualHttpBinding_SampleDuplexHello"
      behaviorConfiguration = "enableCallbackDebug">
    </endpoint>
</client>
<behaviors>
  <endpointBehaviors>
    <behavior name = "enableCallbackDebug">
      <callbackDebug includeExceptionDetailInFaults = "true"/>
    </behavior>
  </endpointBehaviors>
</behaviors>
```

3. 使用 ClientViaBehavior 行为

可以使用 ClientViaBehavior 行为指定应为其创建传输通道的统一资源标识符。当直接网络目标不是消息的预期处理者时,可使用此行为。当调用应用程序不需要知道最终目标时,或者当目标 Via 标头不是地址时,使用此行为可启用多跃点对话。

1.5.5　配置客户端的行为

WCF 通过两种方式配置行为:一是通过引用在客户端应用程序配置文件的 <behavior> 节中定义的行为配置;二是在调用应用程序中采用编程方式进行配置。

1. 以配置文件使用行为

在使用配置文件时,行为配置为配置设置的命名集合。每个行为配置的名称都必须是唯一的。在终结点配置的 behaviorConfiguration 属性中,此字符串用来将终结点链接到该行为。

[示例1]

下面的配置代码定义了一个称为 myBehavior 的行为。客户端终结点在

behaviorConfiguration 属性中引用该行为。

```xml
<configuration>
    <system.serviceModel>
        <behaviors>
            <endpointBehaviors>
                <behavior name = "myBehavior">
                    <clientVia />
                </behavior>
            </endpointBehaviors>
        </behaviors>
        <bindings>
            <basicHttpBinding>
                <binding name = "myBinding" maxReceivedMessageSize = "10000" />
            </basicHttpBinding>
        </bindings>
        <client>
            <endpoint address = "myAddress" binding = "basicHttpBinding" bindingConfiguration = "myBinding" behaviorConfiguration = "myBehavior" contract = "myContract" />
        </client>
    </system.serviceModel>
</configuration>
```

2. 以编程方式使用行为

也可以通过编程方式配置或插入行为,方法是在打开客户端之前查找 Behaviors 客户端对象或客户端的通道工厂对象上的 WCF 属性。

[示例 2]

下面的代码示例演示如何在创建通道对象之前访问从 Behaviors 属性返回的 ServiceEndpoint 上的 Endpoint 属性,从而以编程方式插入行为。

```csharp
public class Client
{   public static void Main()
    {
        try
        {
            ChannelFactory<ISampleServiceChannel> factory
                = new ChannelFactory<ISampleServiceChannel>("WSHttpBinding_
```

```csharp
ISampleService");
        factory.Endpoint.Behaviors.Add(new EndpointBehaviorMessageInspector());
        ISampleServiceChannel wcfClientChannel = factory.CreateChannel();
        Console.WriteLine("Enter the greeting to send: ");
        string greeting = Console.ReadLine();
         Console.WriteLine(" The service responded: " + wcfClientChannel.SampleMethod(greeting));
        Console.WriteLine("Press ENTER to exit:");
        Console.ReadLine();
        wcfClientChannel.Close();
        Console.WriteLine("Done!");
    }
    catch (TimeoutException timeProblem)
    {Console.WriteLine(" The service operation timed out. " + timeProblem.Message);
        Console.Read(); }
    catch (FaultException<SampleFault> fault)
    {    Console.WriteLine("SampleFault fault occurred: {0}", fault.Detail.FaultMessage);
        Console.Read(); }
    catch (CommunicationException commProblem)
    {      Console.WriteLine(" There was a communication problem. " + commProblem.Message);
        Console.Read(); } }
```

第 2 章 服 务 契 约

2.1 单　　向

本节介绍具有单向服务操作的服务协定。客户端不会像在双向服务操作中那样等待服务操作完成。本节基于入门并使用 wsHttpBinding 绑定。本节中的服务是自承载控制台应用程序，通过它可以观察接收和处理请求的服务。客户端也是一个控制台应用程序。

备注

本节最后介绍了此示例的设置过程和生成说明。

若要创建单向服务协定，请定义服务协定，将 OperationContractAttribute 类应用于每个操作，并将 IsOneWay 设置为 true，如下面的示例代码：

```
[ServiceContract(Namespace = "http://Microsoft.ServiceModel.Samples")]
public interface IOneWayCalculator
{
    [OperationContract(IsOneWay = true)]
    void Add(double n1, double n2);
    [OperationContract(IsOneWay = true)]
    void Subtract(double n1, double n2);
    [OperationContract(IsOneWay = true)]
    void Multiply(double n1, double n2);
    [OperationContract(IsOneWay = true)]
    void Divide(double n1, double n2);
}
```

为了演示客户端不会等待服务操作完成，此示例中的服务代码实现了五秒钟的延迟，如下面的示例代码：

```
/This service class implements the service contract.
//This code writes output to the console window.
[ServiceBehavior(ConcurrencyMode = ConcurrencyMode.Multiple,
    InstanceContextMode = InstanceContextMode.PerCall)]
```

```csharp
public class CalculatorService : IOneWayCalculator
{
    public void Add(double n1, double n2)
    {
        Console.WriteLine("Received Add({0},{1}) - sleeping", n1, n2);
        System.Threading.Thread.Sleep(1000 * 5);
        double result = n1 + n2;
        Console.WriteLine("Processing Add({0},{1}) - result:{2}", n1, n2, result);
    }
    ...}
```

当客户端调用服务时,调用不等待服务操作完成即返回。

运行示例时,客户端和服务活动将显示在服务和客户端控制台窗口中。可以看到服务从客户端接收消息。在每个控制台窗口中按 Enter 可以同时关闭服务和客户端。

客户端在服务之前完成,说明了客户端没有等待单向服务操作完成。客户端输出如下:

```
Add(100,15.99)
Subtract(145,76.54)
Multiply(9,81.25)
Divide(22,7)

Press <ENTER> to terminate client.
```

服务输出如下:

```
The service is ready.
Press <ENTER> to terminate service.
Received Add(100,15.99) - sleeping
Received Subtract(145,76.54) - sleeping
Received Multiply(9,81.25) - sleeping
Received Divide(22,7) - sleeping
Processing Add(100,15.99) - result:115.99
Processing Subtract(145,76.54) - result:68.46
Processing Multiply(9,81.25) - result:731.25
Processing Divide(22,7) - result:3.14285714285714
```

备注

HTTP 从定义上讲是一个请求/响应协议;当发出请求时,即返回响应。即使对于通过 HTTP 公开的单向服务操作,也是如此。当调用操作时,服务在执行服务操作之前返回 HTTP 状态码 202。此状态码表示请求已被接受进行处理,但处理尚未完成。调用操作的客户端在从服务收到 202 响应之前处于阻止状态。当使用绑定(配置为使用会话)发送多个单向消息时,这可能会产生某些意外行为。此示例中使用的 wsHttpBinding 绑定配置为默认使用会话来建立安全上下文。默认情况下,会话中的消息一定会按照它们的发送顺序到达,因此当发送会话中的第二个消息时,在处理完第一个消息之前不会处理第二个消息。这样的结果是,在处理完上一个消息之前,客户端不会收到消息的 202 响应,因此客户端似乎是阻止了每个后续的操作调用。为了避免此行为,此示例对运行库进行了配置,以便将消息并发调度给不同的实例进行处理。本示例将 InstanceContextMode 设置为 PerCall,以使每条消息可以由不同的实例来处理。ConcurrencyMode 设置为 Multiple,以允许多个线程同时调度消息。

设置、生成和运行示例

确保已执行的 Windows Communication Foundation 示例的一次性安装过程。

若要生成 C# 或 Visual Basic .NET 版本的解决方案,请按照 Building the Windows Communication Foundation Samples 中的说明进行操作。

若要在单或跨计算机配置上运行示例,请按上面的说明运行 Windows Communication Foundation 示例。

备注

请在运行客户端之前运行服务,并在关闭服务之前关闭客户端。这样可以避免当客户端由于服务已关闭而无法彻底关闭安全会话时出现客户端异常。

重要

计算机上可能已安装这些示例。在继续操作之前,请先检查以下(默认)目录:

<InstallDrive>:\WF_WCF_Samples

如果此目录不存在,请访问 针对 .NET Framework 4 的 Windows Communication Foundation(WCF)和 Windows Workflow Foundation(WF)示例 以下载所有 Windows Communication Foundation(WCF)和 WF 示例。此示例位于以下目录:

<InstallDrive>:\WF_WCF_Samples\WCF\Basic\Contract\Service\Oneway

2.2 双 工

本节介绍如何定义和实现双工协定。当客户端与服务建立会话并为服务提供可用来将消息发送回客户端的通道时,就会发生双工通信。此示例基于入门。双工协定以一对接口形式定义:一个从客户端到服务端的主接口和一个从服务端到客户端的回调接口。在本节中,ICalculatorDuplex 接口允许客户端执行数学运算,通过会话计算结果。服务在 ICalculatorDuplexCallback 接口上返回结果。双工协定需要会话,因为必须建立上下文才能将客户端和服务之间发送的一组消息关联在一起。

备注

本节最后介绍了此示例的设置过程和生成说明。

在此示例中,客户端是一个控制台应用程序(.exe),服务是由 Internet 信息服务(IIS)承载的。双工协定定义如下:

```
[ServiceContract(Namespace = "http://Microsoft.ServiceModel.Samples", SessionMode = SessionMode.Required, CallbackContract = typeof(ICalculatorDuplexCallback))]
public interface ICalculatorDuplex
{
    [OperationContract(IsOneWay = true)]
    void Clear();
    [OperationContract(IsOneWay = true)]
    void AddTo(double n);
    [OperationContract(IsOneWay = true)]
    void SubtractFrom(double n);
    [OperationContract(IsOneWay = true)]
    void MultiplyBy(double n);
    [OperationContract(IsOneWay = true)]
    void DivideBy(double n);
}

public interface ICalculatorDuplexCallback
{
    [OperationContract(IsOneWay = true)]
    void Result(double result);
    [OperationContract(IsOneWay = true)]
    void Equation(string eqn);
}
```

CalculatorService 类实现 ICalculatorDuplex 主接口。服务使用 PerSession 实例模式来维护每个会话的结果。名为 Callback 的私有属性用于访问指向客户端的回调通道。服务使用该回调通过回调接口将消息发送回客户端。

```
[ServiceBehavior(InstanceContextMode = InstanceContextMode.PerSession)]
public class CalculatorService : ICalculatorDuplex
{
    double result = 0.0D;
    string equation;
```

```
public CalculatorService()
{
    equation = result.ToString();
}

public void Clear()
{
    Callback.Equation(equation + " = " + result.ToString());
    equation = result.ToString();
}

public void AddTo(double n)
{
    result += n;
    equation += " + " + n.ToString();
    Callback.Result(result);
}
...
ICalculatorDuplexCallback Callback
{
    get
    {
        return OperationContext.Current.GetCallbackChannel<ICalculatorDuplexCallback>();
    }
}
}
```

客户端必须提供一个实现双工协定回调接口的类,以便从服务接收消息。该示例中,定义 CallbackHandler 类以实现 ICalculatorDuplexCallback 接口。

```
public class CallbackHandler : ICalculatorDuplexCallback
{
    public void Result(double result)
    {
        Console.WriteLine("Result({0})", result);
    }

    public void Equation(string equation)
```

```
        }
        Console.WriteLine("Equation({0}", equation);
    }
}
```

针对双工协定生成的代理需要在构造时提供一个 InstanceContext。此 InstanceContext 用作实现回调接口并处理从服务发送回的消息的对象所在的位置。InstanceContext 是用 CallbackHandler 类的实例构造的。此对象处理通过回调接口从服务发送到客户端的消息。

```
//Construct InstanceContext to handle messages on callback interface.
InstanceContext instanceContext = new InstanceContext(new CallbackHandler());

//Create a client.
CalculatorDuplexClient client = new CalculatorDuplexClient(instanceContext);

Console.WriteLine(" Press <ENTER> to terminate client once the output is displayed.");
Console.WriteLine();

//Call the AddTo service operation.
double value = 100.00D;
client.AddTo(value);

//Call the SubtractFrom service operation.
value = 50.00D;
client.SubtractFrom(value);

//Call the MultiplyBy service operation.
value = 17.65D;
client.MultiplyBy(value);

//Call the DivideBy service operation.
value = 2.00D;
client.DivideBy(value);

//Complete equation.
client.Clear();
```

```
Console.ReadLine();

//Closing the client gracefully closes the connection and cleans up resources.
client.Close();
```

已将配置修改为提供同时支持会话通信和双工通信的绑定。wsDualHttpBinding 支持会话通信,并通过提供双 HTTP 连接(一个连接对应于一个方向)来允许进行双工通信。在服务上,配置中的唯一区别是所用的绑定。在客户端上,必须配置一个服务器可用来连接客户端的地址,如下面的示例配置所示。

```
<client>
  <endpoint name = ""
           address = "http://localhost/servicemodelsamples/service.svc"
           binding = "wsDualHttpBinding"
           bindingConfiguration = "DuplexBinding"
           contract = "Microsoft.ServiceModel.Samples.ICalculatorDuplex" />
</client>

<bindings>
  <!-- Configure a binding that support duplex communication. -->
  <wsDualHttpBinding>
    <binding name = "DuplexBinding"
            clientBaseAddress = "http://localhost:8000/myClient/" >
    </binding>
  </wsDualHttpBinding>
</bindings>
```

当运行示例时,从服务发送的回调接口上,可以看到返回到客户端的消息。显示每个中间结果,最后在所有操作完成时显示整个公式。按 Enter 可关闭客户端。

设置、生成和运行示例

确保已执行的 Windows Communication Foundation 示例的一次性安装过程。

若要生成解决方案的 C#、C++ 或 Visual Basic.NET 版本,请按上述示例的说明生成 Windows Communication Foundation 示例。

若要在单或跨计算机配置上运行示例,请按照上述示例的说明运行 Windows Communication Foundation 示例。

重要

(1)在运行时客户端的跨计算机配置,请务必将中的"localhost" address 属性终结点元素和 clientBaseAddress 属性 <绑定> 元素 <wsDualHttpBinding> 具有适当的计算机,如以

下所示的名称的元素:
< client >
< endpoint name = " "
address = "http://service_machine_name/servicemodelsamples/service.svc"
... />
< /client >
...
< wsDualHttpBinding >
< binding name = " DuplexBinding" clientBaseAddress = " http://client_machine_name:8000/myClient/" >
< /binding >
< /wsDualHttpBinding >

（2）您计算机上可能已安装这些示例。在继续操作之前,请先检查以下（默认）目录：
< InstallDrive > :\WF_WCF_Samples

如果此目录不存在,请访问 针对 .NET Framework 4 的 Windows Communication Foundation（WCF）和 Windows Workflow Foundation（WF）示例 以下载所有 Windows Communication Foundation（WCF）和 WF 示例。此示例位于以下目录：

< InstallDrive > :\WF_WCF_Samples\WCF\Basic\Contract\Service\Duplex

2.3 错 误 协 定

"错误协定"演示如何将错误信息从服务传达到客户端。本节基于入门,使用的一些其他代码添加到服务,以将内部异常转换为错误。客户端试图执行除数为零的运算以在服务上强制产生错误情况。

备注

本节的最后介绍了此示例的设置过程和生成说明。

已将计算器协定修改为包括 FaultContractAttribute,如下面的示例代码所示。
[ServiceContract(Namespace = "http://Microsoft.ServiceModel.Samples")]
public interface ICalculator
{
　　[OperationContract]
　　int Add(int n1, int n2);
　　[OperationContract]
　　int Subtract(int n1, int n2);

```
    [OperationContract]
    int Multiply(int n1, int n2);
    [OperationContract]
    [FaultContract(typeof(MathFault))]
    int Divide(int n1, int n2);
}
```

FaultContractAttribute 属性指示 Divide 运算可能返回一个 MathFault 类型的错误。错误可以是可序列化的任何类型。在本例中，MathFault 为数据协定如下：

```
[DataContract(Namespace = "http://Microsoft.ServiceModel.Samples")]
public class MathFault
{
    private string operation;
    private string problemType;

    [DataMember]
    public string Operation
    {
        get { return operation; }
        set { operation = value; }
    }

    [DataMember]
    public string ProblemType
    {
        get { return problemType; }
        set { problemType = value; }
    }
}
```

如下面的示例代码所示，发生除数为零异常时，Divide 方法引发 FaultException<TDetail>异常。此异常导致向客户端发送一个错误。

```
public int Divide(int n1, int n2)
{
    try
    {
        return n1 /n2;
    }
    catch (DivideByZeroException)
```

```
        MathFault mf = new MathFault();
        mf.operation = "division";
        mf.problemType = "divide by zero";
        throw new FaultException<MathFault>(mf);
    }
}
```

通过请求除数为零,客户端代码强制产生错误。运行示例时,操作请求和响应将显示在客户端控制台窗口中。您会看到将除数为零报告为错误。在客户端窗口中按 Enter 可以关闭客户端。

```
Add(15,3) = 18
Subtract(145,76) = 69
Multiply(9,81) = 729
FaultException<MathFault>: Math fault while doing division. Problem: divide by zero

Press <ENTER> to terminate client.
```

客户端通过捕获相应 FaultException<MathFault> 异常来完成此操作:

```
catch(FaultException<MathFault> e)
{
    Console.WriteLine("FaultException<MathFault>: Math fault while doing "
+ e.Detail.operation + ". Problem: " + e.Detail.problemType);
    client.Abort();
}
```

默认情况下,不可预期的异常的详细信息不发送到客户端,以防止服务实现的详细信息泄露出服务的安全边界。FaultContract 提供一种方法来描述协定中的故障并将某些类型的异常标记为适合传输到客户端。FaultException<T> 提供了用于向使用者发送故障的运行时机制。

但是,在调试时查看服务失败的内部详细信息很有用。若要关闭上述安全行为,可以指示在发送到客户端的错误中必须包括服务器上每个未处理的异常的详细信息。这是通过将 IncludeExceptionDetailInFaults 设置为 true 来完成的。可以在代码中或在配置中设置它,如下面的示例所示。

```
<behaviors>
  <serviceBehaviors>
    <behavior name="CalculatorServiceBehavior">
      <serviceMetadata httpGetEnabled="True"/>
```

```
      <serviceDebug includeExceptionDetailInFaults = "True" />
    </behavior>
  </serviceBehaviors>
</behaviors>
```

此外,行为必须将配置文件中的 behaviorConfiguration 属性设置为"Calculator Service Behavior"。

若要在客户端上捕获这样的错误,必须捕获非泛型 FaultException。

此行为仅用于调试目的,切勿在生产中启用它。

设置、生成和运行示例

(1)确保已执行的 Windows Communication Foundation 示例的一次性安装过程。

(2)若要生成 C# 或 Visual Basic .NET 版本的解决方案,请按照 Building the Windows Communication Foundation Samples 中的说明进行操作。

(3)若要在单或跨计算机配置上运行示例,请按第(1)步的操作说明运行 Windows Communication Foundation 示例。

重要

计算机上可能已安装这些示例。在继续操作之前,请先检查以下(默认)目录:

<InstallDrive>:\WF_WCF_Samples

如果此目录不存在,请访问 针对 .NET Framework 4 的 Windows Communication Foundation (WCF) 和 Windows Workflow Foundation (WF) 示例 以下载所有 Windows Communication Foundation (WCF) 和 WF 示例。此示例位于以下目录:

<InstallDrive>:\WF_WCF_Samples\WCF\Basic\Contract\Service\Faults

2.4 会 话

"会话"示例演示如何实现需要会话的协定。会话用来提供执行多个操作的上下文。这允许服务将某个状态与给定的会话相关联,从而使后续操作可以使用上一个操作的状态。此示例基于入门,该类可实现计算器服务。ICalculator 协定已进行修改,允许在保持运行结果的同时执行一组算术运算。此功能由 ICalculatorSession 协定定义。服务在调用多个服务操作以执行客户端运行时的状态。客户端可以通过调用 Result() 来检索当前结果,通过调用 Clear() 将结果清零。

在本节中,客户端是一个控制台应用程序(.exe),服务是由 Internet 信息服务(IIS)承载的。

备注

本主题的最后介绍了此示例的设置过程和生成说明。

将协定的 SessionMode 设置为 Required 可以确保在通过特定的绑定公开该协定时,该绑定支持会话。如果绑定不支持会话,则会引发异常。定义 ICalculatorSession 接口使得可以调用一个或多个修改运行结果的操作,如下面的示例代码所示。

```
[ServiceContract(Namespace = " http://Microsoft.ServiceModel.Samples",
SessionMode = SessionMode.Required)]
public interface ICalculatorSession
{
    [OperationContract(IsOneWay = true)]
    void Clear();
    [OperationContract(IsOneWay = true)]
    void AddTo(double n);
    [OperationContract(IsOneWay = true)]
    void SubtractFrom(double n);
    [OperationContract(IsOneWay = true)]
    void MultiplyBy(double n);
    [OperationContract(IsOneWay = true)]
    void DivideBy(double n);
    [OperationContract]
    double Result();
}
```

服务使用 InstanceContextMode 的 PerSession 将给定的服务实例上下文绑定到每个传入会话。这使服务可以在本地成员变量中维护每个会话的运行结果。

```
[ServiceBehavior(InstanceContextMode = InstanceContextMode.PerSession)]
public class CalculatorService : ICalculatorSession
{
    double result = 0.0D;

    public void Clear()
    { result = 0.0D; }

    public void AddTo(double n)
    { result += n; }

    public void SubtractFrom(double n)
```

```
    result - = n; }

    public void MultiplyBy(double n)
    { result * = n; }

    public void DivideBy(double n)
    { result /= n; }

    public double Result()
    { return result; }
}
```

运行示例时，客户端会向服务器发出几个请求并请求结果，结果显示在客户端控制台窗口中。在客户端窗口中按 Enter 可以关闭客户端。

```
(((0 + 100) - 50) * 17.65) /2 = 441.25
Press <ENTER> to terminate client.
```

设置、生成和运行示例

(1) 确保已执行的 Windows Communication Foundation 示例的一次性安装过程。

(2) 若要生成 C# 或 Visual Basic .NET 版本的解决方案，请按照 Building the Windows Communication Foundation Samples 中的说明进行操作。

(3) 若要在单或跨计算机配置上运行示例，请按照中的说明运行 Windows Communication Foundation 示例。

重要

计算机上可能已安装这些示例。在继续操作之前，请先检查以下(默认)目录：

<InstallDrive>:\WF_WCF_Samples

如果此目录不存在，请访问 针对 .NET Framework 4 的 Windows Communication Foundation (WCF) 和 Windows Workflow Foundation (WF) 示例 以下载所有 Windows Communication Foundation (WCF) 和 WF 示例。此示例位于以下目录：

<InstallDrive>:\WF_WCF_Samples\WCF\Basic\Contract\Service\Session

2.5 流

流示例演示流传输模式通信的用法。收发流的若干操作都由服务公开。本节讲述自承载主题。客户端和服务都是控制台程序。

Windows Communication Foundation（WCF）可以在两个传输模式下通信：缓冲模式和流模式。在默认的缓冲传输模式中，必须完整传递消息，接收方才能读取该消息。在流传输模式下，接收方可以在完整传递消息之前开始处理该消息。当传递的信息很长且可以依次处理时，流模式非常有用。当消息过长以致无法全部缓冲时，流模式也非常有用。

1. 流处理和服务协定

在设计服务协定时，可以考虑进行流处理。如果某个操作接收或返回大量数据，则应当考虑对该数据进行流处理，以免因对输入或输出消息进行缓冲而导致内存使用率过高。若要对数据进行流处理，用来存放该数据的参数必须是消息中的唯一参数。例如，如果要对输入消息进行流处理，则该操作必须正好具有一个输入参数。同样，如果要对输出消息进行流处理，则该操作必须正好具有一个输出参数或一个返回值。在任一情况下，该参数或返回值类型必须是 Stream、Message 或 IXmlSerializable。下面是该流处理示例中使用的服务协定。

```
[ServiceContract(Namespace = "http://Microsoft.ServiceModel.Samples")]
public interface IStreamingSample
{
    [OperationContract]
    Stream GetStream(string data);
    [OperationContract]
    bool UploadStream(Stream stream);
    [OperationContract]
    Stream EchoStream(Stream stream);
    [OperationContract]
    Stream GetReversedStream();
}
```

GetStream 操作接收一些输入数据作为经过缓冲的字符串，并返回经过流处理的 Stream。相反，UploadStream 接收一个经过流处理的 Stream，并返回一个经过缓冲的 bool。EchoStream 是输入消息和输出消息都经过流处理的操作的示例，它接收和返回 Stream。最后，GetReversedStream 将不接收输入，并返回一个 Stream（已经过流处理）。

2. 启用流传输

按照上面的说明定义操作协定将在编程模型级别提供流处理。如果仅限于此，则传输仍将缓冲整个消息内容。若要启用传输流，需要在传输的绑定元素上选择传输模式。该绑定元素具有一个 TransferMode 属性，该属性可以设置为 Buffered、Streamed、StreamedRequest 或 StreamedResponse。将传输模式设置为 Streamed，将在两个方向上启用流通信。将传输模式设置为 StreamedRequest 或 StreamedResponse 将分别只在请求或响应中启用流通信。

basicHttpBinding 公开绑定上的 TransferMode 属性，这与 NetTcpBinding 和

NetNamedPipeBinding 一样。对于其他传输,必须创建一个自定义绑定以设置传输模式。

如下示例中的配置代码演示如何将 TransferMode 和自定义 HTTP 绑定上的 basicHttpBinding 属性设置为流处理:

```xml
<!-- An example basicHttpBinding using streaming. -->
<basicHttpBinding>
  <binding name = "HttpStreaming" maxReceivedMessageSize = "67108864"
        transferMode = "Streamed"/>
</basicHttpBinding>
<!-- An example customBinding using HTTP and streaming. -->
<customBinding>
  <binding name = "Soap12">
    <textMessageEncoding messageVersion = "Soap12WSAddressing10"/>
    <httpTransport transferMode = "Streamed"
            maxReceivedMessageSize = "67108864"/>
  </binding>
</customBinding>
```

除了将 transferMode 设置为 Streamed 以外,上面的配置代码还将 maxReceivedMessageSize 设置为 64 MB。作为一个防范机制,maxReceivedMessageSize 为允许接收的最大消息设置了上限。默认的 maxReceivedMessageSize 为 64 KB,对于流处理方案而言,此值通常太低。

3. 处理经过流处理的数据

GetStream、UploadStream 和 EchoStream 操作都可以处理直接从文件发送数据或将接收的数据直接保存到文件的情况。但是,在某些情况下,需要发送或接收大量数据,并对正在发送或接收的数据块执行某种处理。解决类似方案的一种方法是编写一个自定义流(一个派生自 Stream 的类)来处理读取或写入的数据。GetReversedStream 操作和 ReverseStream 类就是类似方法的一个示例。

GetReversedStream 创建并返回 ReverseStream 的新实例。当系统从该 ReverseStream 对象中进行读取时,会发生实际的处理。ReverseStream.Read 实现从基础文件中读取字节块区,反转字节,然后返回经过反转的字节。这不会反转整个文件内容,而是一次仅反转一个字节块区。下面的示例演示在流中读取或写入内容时如何进行流处理。

```csharp
class ReverseStream : Stream
{
    FileStream inStream;
    internal ReverseStream(string filePath)
    {
```

```
        //Opens the file and places a StreamReader around it.
        inStream = File.OpenRead(filePath);
    }
    //Other methods removed for brevity.
    public override int Read(byte[] buffer, int offset, int count)
    {
        int countRead = inStream.Read(buffer, offset, count);
        ReverseBuffer(buffer, offset, countRead);
        return countRead;
    }
    public override void Close()
    {
        inStream.Close();
        base.Close();
    }
    protected override void Dispose(bool disposing)
    {
        inStream.Dispose();
        base.Dispose(disposing);
    }
    void ReverseBuffer(byte[] buffer, int offset, int count)
    {
        int i, j;
        for (i = offset, j = offset + count - 1; i < j; i++, j--)
        {
            byte currenti = buffer[i];
            buffer[i] = buffer[j];
            buffer[j] = currenti;
        }
    }
}
```

4. 运行示例

若要运行此示例,请首先按照本文档末尾的说明生成服务和客户端,然后在两个不同的控制台窗口中启动服务和客户端。当客户端启动时,它会在服务就绪时等待用户按 Enter。客户端随后会依次通过 HTTP 和 TCP 来调用 GetStream( )、UploadStream( ) 和 GetReversedStream( ) 方法。下面是服务的示例输出,其后面是客户端的示例输出。

服务输出：
```
The streaming service is ready.
Press <ENTER> to terminate service.


Saving to file D:\...\uploadedfile
....................
File D:\...\uploadedfile saved
Saving to file D:\...\uploadedfile
..............
File D:\...\uploadedfile saved
```
客户端输出：
```
Press <ENTER> when service is ready
------ Using HTTP ------
Calling GetStream()
Saving to file D:\...\clientfile
....................
Wrote 33405 bytes to stream

File D:\...\clientfile saved
Calling UploadStream()
Calling GetReversedStream()
Saving to file D:\...\clientfile
....................
Wrote 33405 bytes to stream

File D:\...\clientfile saved
------ Using Custom HTTP ------
Calling GetStream()
Saving to file D:\...\clientfile
..............
Wrote 33405 bytes to stream

File D:\...\clientfile saved
Calling UploadStream()
Calling GetReversedStream()
Saving to file D:\...\clientfile
..............
```

```
Wrote 33405 bytes to stream

File D:\...\clientfile saved

Press <ENTER> to terminate client.
```

5. 设置、生成和运行示例

确保已执行的 Windows Communication Foundation 示例的一次性安装过程。

若要生成 C# 或 Visual Basic .NET 版本的解决方案,请按照 Building the Windows Communication Foundation Samples 中的说明进行操作。

若要在单或跨计算机配置上运行示例,请按照说明运行 Windows Communication Foundation 示例。

备注

如果使用 Svcutil.exe 为此示例重新生成配置,请确保在客户端配置中修改终结点名称以与客户端代码匹配。

重要

计算机上可能已安装这些示例。在继续操作之前,请先检查以下(默认)目录:

`<InstallDrive>:\WF_WCF_Samples`

如果此目录不存在,请访问 针对 .NET Framework 4 的 Windows Communication Foundation (WCF) 和 Windows Workflow Foundation (WF) 示例 以下载所有 Windows Communication Foundation (WCF) 和 WF 示例。此示例位于以下目录:

`<InstallDrive>:\WF_WCF_Samples\WCF\Basic\Contract\Service\Stream`

第 3 章 数据协定

3.1 数据协定基本知识

3.1.1 数据协定概述

"数据协定"是在服务与客户端之间达成的正式协议,用于以抽象方式描述要交换的数据。也就是说,为了进行通信,客户端和服务不必共享相同的类型,而只需共享相同的数据协定。数据协定为每个参数或返回类型精确定义为进行交换而序列化哪些数据(将哪些数据转换为 XML)。

1. 数据协定基本知识

默认情况下,Windows Communication Foundation(WCF)使用称为数据协定序列化程序的序列化引擎对数据进行序列化和反序列化(与 XML 进行相互转换)。所有 .NET Framework 基元类型(如整型和字符串型)以及某些被视为基元的类型(如 DateTime 和 XmlElement)无须做其他任何准备工作就可序列化并被视为拥有默认数据协定。许多 .NET Framework 类型也具有现有数据协定。

必须为所创建的新复杂类型定义数据协定才能序列化这些类型。默认情况下,DataContractSerializer 推断数据协定并序列化所有公共可见类型。类型的所有公共读/写属性和字段均被序列化。可以使用 IgnoreDataMemberAttribute 从序列化中剔除某些成员。还可以使用 DataContractAttribute 和 DataMemberAttribute 属性显式创建数据协定。正常情况下可通过将 DataContractAttribute 属性应用到该类型来完成该任务。可以将此属性应用到类、结构和枚举。然后必须将 DataMemberAttribute 属性应用到数据协定类型的每个成员,以指示这些成员为数据成员,即应进行序列化。

下面的示例显式应用了 ServiceContractAttribute 和 OperationContractAttribute 属性的服务协定(接口)。该示例演示基元类型不要求数据协定,而复杂类型却对此有要求。

```
[ServiceContract]
public interface ISampleInterface
```

```
    //No data contract is required since both the parameter
    //and return types are primitive types.
    [OperationContract]
    double SquareRoot(int root);
    //No Data Contract required because both parameter and return
    //types are marked with the SerializableAttribute attribute.
    [OperationContract]
    System.Drawing.Bitmap GetPicture(System.Uri pictureUri);
    //The MyTypes.PurchaseOrder is a complex type, and thus
    //requires a data contract.
    [OperationContract]
    bool ApprovePurchaseOrder(MyTypes.PurchaseOrder po);
}
```

下面的示例演示如何通过将 MyTypes.PurchaseOrder 和 DataContractAttribute 属性应用于类及其成员来创建 DataMemberAttribute 类型的数据协定。

```
namespace MyTypes
{
    [DataContract]
    public class PurchaseOrder
    {
        private int poId_value;
        //Apply the DataMemberAttribute to the property.
        [DataMember]
        public int PurchaseOrderId
        {
            get { return poId_value; }
            set { poId_value = value; }
        }
    }
}
```

2. 说明

下面的注释提供在创建数据协定时需要考虑的事项：

仅当用于未标记的类型时，才接受 IgnoreDataMemberAttribute 属性。这包括未使用 DataContractAttribute、SerializableAttribute、CollectionDataContractAttribute 或 EnumMemberAttribute 属性之一标记的类型，或通过任何其他方式（如 IXmlSerializable）标记为可序列化的类型。

可以将 DataMemberAttribute 属性（Attribute）应用于字段和属性（Property）。

成员可访问性级别（internal、private、protected 或 public）对数据协定无任何影响。

如果将 DataMemberAttribute 属性应用于静态成员，则将忽略该属性。

在序列化期间，为属性数据成员调用 property-get 代码来获取要序列化的属性的值。

在反序列化期间，首先创建一个未初始化的对象，而不在该类型上调用任何构造函数。然后反序列化所有数据成员。

在反序列化期间，为属性数据成员调用 property-set 代码，将属性设置为要反序列化的值。

对于将要生效的数据协定，它必须能序列化其所有数据成员。

泛型类型的处理方式与非泛型类型完全相同。泛型参数无特殊要求。例如，请注意以下类型：

```
[DataContract]
public class MyGenericType1<T>
{
    //Code not shown.
}
```

无论用于泛型类型参数（T）的类型能否序列化，此类型都可序列化。因为它必须能序列化所有数据成员，所以下面的类型仅在泛型类型参数也可序列化时才可序列化，如以下代码所示：

```
[DataContract]
public class MyGenericType2<T>
{
    [DataMember]
    T theData;
}
```

3.1.2 创建类或结构的基本数据协定

1. 创建类或结构的基本数据协定

通过将 DataContractAttribute 属性应用于类来声明该类型具有数据协定。请注意，包括不带属性的公共类型在内的所有公共类型都是可序列化的。如果不存在 DataContractSerializer 属性，DataContractAttribute 将推断出一个数据协定。有关详细信息，请参见可序列化类型。

通过将 DataMemberAttribute 属性（Attribute）应用于每个成员来定义要序列化的成员（属性（Property）、字段或事件）。这些成员称为数据成员。默认情况下，所有公共类型都是可序列化的。

备注

可以将 DataMemberAttribute 属性应用于私有字段,这会导致向其他人公开此数据。请确保成员不包含敏感数据。

2. 示例

下面的示例演示如何通过将 Person 和 DataContractAttribute 属性应用于类及其成员来创建 DataMemberAttribute 类型的数据协定。

```
using System;
using System.Runtime.Serialization;
[DataContract]
public class Person
{
    //This member is serialized.
    [DataMember]
    internal string FullName;
    //This is serialized even though it is private.
    [DataMember]
    private int Age;
    //This is not serialized because the DataMemberAttribute
    //has not been applied.
    private string MailingAddress;
    //This is not serialized, but the property is.
    private string telephoneNumberValue;
    [DataMember]
    public string TelephoneNumber
    {
        get { return telephoneNumberValue; }
        set { telephoneNumberValue = value; }
    }
}
```

3.2 可序列化类型

默认情况下,DataContractSerializer 序列化所有公共可见类型。类型的所有公共读/写属性和字段均被序列化。

可以通过对类型和成员应用 DataContractAttribute 和 DataMemberAttribute 属性来更改默认行为。当用户拥有不受用户控制因而无法对其进行修改以添加属性的类型时,此功能可能会十分有用。DataContractSerializer 可识别这类"未标记"的类型。

1. 序列化默认设置

可以应用 DataContractAttribute 和 DataMemberAttribute 属性以显式控制或自定义类型和成员的序列化。此外,还可以将这些属性应用于私有字段。但是,即使未用这些属性进行标记的类型也会进行序列化和反序列化。适用的规则和例外如下:

- DataContractSerializer 使用新创建的类型的默认属性从不带属性的类型推断数据协定。
- 除非对成员应用 get 属性(Attribute),否则所有公共字段以及具有公共 set 和 IgnoreDataMemberAttribute 方法的属性(Property)都会序列化。
- 序列化语义与 XmlSerializer 的语义类似。
- 在未标记的类型中,仅序列化具有不带参数的构造函数的公共类型。此规则的例外是用于 ExtensionDataObject 接口的 IExtensibleDataObject。
- 只读字段、没有 get 或 set 方法的属性,以及具有内部或私有 set 或 get 方法的属性不会进行序列化。此类属性会被忽略,但不会引发异常(get – only 集合的情况除外)。
- 会忽略 XmlSerializer 属性(如 XmlElement、XmlAttribute、XmlIgnore、XmlInclude 等)。
- 如果未将 DataContractAttribute 属性应用于某个给定类型,则序列化程序会忽略该类型中应用了 DataMemberAttribute 属性的所有成员。
- 未使用 KnownTypes 属性(Attribute)进行标记的类型中支持 DataContractAttribute 属性(Property)。这包括对未标记类型上的 KnownTypeAttribute 属性(Attribute)的支持。
- 若要使公共成员、属性(Property)或字段"退出"序列化过程,请向该成员应用 IgnoreDataMemberAttribute 属性(Attribute)。

2. 继承

未标记类型(没有 DataContractAttribute 属性的类型)可以从具有此属性的类型继承;但是反过来则不允许:具有该属性的类型不能从未标记类型继承。实行此规则主要是为了确保与使用旧版本 .NET Framework 编写的代码向后兼容。

3.3 数据成员顺序

在一些应用程序中,有必要知道各个数据成员中数据的发送顺序或预期接收顺序(比如序列化 XML 中数据的显示顺序)。有时,必须要更改此顺序。本节说明排序规则。

1. 基本规则

数据排序的基本规则包括:

如果数据协定类型是继承层次结构的一部分,则其基类型的数据成员始终排在第一位。

排在下一位的是当前类型的数据成员(按字母顺序排列),这些成员未设置 Order 属性(Attribute)的 DataMemberAttribute 属性(Property)。

再下面是设置了 Order 属性(Attribute)的 DataMemberAttribute 属性(Property)的任何数据成员。这些成员首先按 Order 属性的值排序,如果多个成员具有特定的 Order 值,则按字母顺序排列。可以跳过 Order 值。

字母顺序是通过调用 CompareOrdinal 方法确定的。

2. 示例

考虑下列代码。

```
[DataContract]
public class BaseType
{

    [DataMember]
    public string zebra;
}
[DataContract]
public class DerivedType : BaseType
{

    [DataMember(Order = 0)]
    public string bird;
```

```
    [DataMember(Order = 1)]
    public string parrot;
    [DataMember]
    public string dog;
    [DataMember(Order = 3)]
    public string antelope;
    [DataMember]
    public string cat;
    [DataMember(Order = 1)]
    public string albatross;
}
```

生成的 XML 类似于以下形式。

```
<DerivedType>
    <!-- Zebra is a base data member, and appears first. -->
    <zebra/>
    <!-- Cat has no Order, appears alphabetically first. -->
    <cat/>
    <!-- Dog has no Order, appears alphabetically last. -->
    <dog/>
    <!-- Bird is the member with the smallest Order value -->
    <bird/>
    <!-- Albatross has the next Order value, alphabetically first. -->
    <albatross/>
    <!-- Parrot, with the next Order value, alphabetically last. -->
    <parrot/>
    <!-- Antelope is the member with the highest Order value. Note that Order=2 is skipped -->
    <antelope/>
</DerivedType>
```

3.4 数据协定已知类型

KnownTypeAttribute 类允许用户预先指定应该在反序列化期间包括在考虑范围内的类型。

通常,在客户端和服务之间传递参数和返回值时,这两个终结点共享要传输的数据的所有数据协定。但是,在以下情况下并非如此:

◇ 已发送的数据协定派生自预期的数据协定。在该情况下,传输的数据没有与接收终结点所预期相同的数据协定。

◇ 要传输的信息的声明类型是接口,而非类、结构或枚举。因此,无法预先知道实际发送了实现接口的哪个类型,接收终结点就无法预先确定已传输数据的数据协定。

◇ 要传输的信息的声明类型是 Object。由于每个类型都继承自 Object,而且无法预先知道实际发送了哪个类型,因此接收终结点无法预先确定已传输数据的数据协定。这是第一个项的特殊情况:每个数据协定都源自为 Object 生成的默认空数据协定。

◇ 某些类型(包括 .NET Framework 类型)具有上述三种类别之一中的成员。例如,Hashtable 使用 Object 在哈希表中存储实际对象。在序列化这些类型时,接收方无法预先确定这些成员的数据协定。

1. KnownTypeAttribute 类

在数据到达接收终结点时,WCF 运行库尝试将数据反序列化为公共语言运行库(CLR)类型的实例。通过首先检查传入消息选择为反序列化而实例化的类型,以确定消息内容遵循的数据协定。然后反序列化引擎尝试查找实现与消息内容兼容的数据协定的 CLR 类型。反序列化引擎在此过程中允许的候选类型集称为反序列化程序的"已知类型"集。

让反序列化引擎了解某个类型的一种方法是使用 KnownTypeAttribute。不能将属性应用于单个数据成员,只能将它应用于整个数据协定类型。将属性应用于可能为类或结构的"外部类型"。在其最基本的用法中,应用属性会将类型指定为"已知类型"。只要反序列化外部类型的对象或通过其成员引用的任何对象,这就会导致已知类型成为已知类型集的一部分。可以将多个 KnownTypeAttribute 属性应用于同一类型。

2. 已知类型和基元

基元类型以及被视为基元的某些类型(例如 DateTime 和 XmlElement)始终是"已知"的,且从来不必通过此机制进行添加。但是,必须显式添加基元类型的数组。大多数集合被视为等效于数组(非泛型集合被视为等效于 Object 的数组)。

备注

与其他基元类型不同,DateTimeOffset 结构默认情况下不是已知类型,因此必须将它手动添加到已知类型列表。

3. 示例

下面的示例说明如何使用 KnownTypeAttribute 类。

[示例 1]

有三个具有继承关系的类。

```
[DataContract]
public class Shape { }
[DataContract(Name = "Circle")]
public class CircleType : Shape { }
[DataContract(Name = "Triangle")]
public class TriangleType : Shape { }
```

如果 CompanyLogo 成员设置为 ShapeOfLogo 或 CircleType 对象,则可以序列化下面的 TriangleType 类,而不能对其进行反序列化,因为反序列化引擎无法识别具有数据协定名称 "Circle" 或 "Triangle" 的任何类型。

```
[DataContract]
public class CompanyLogo
{
    [DataMember]
    private Shape ShapeOfLogo;
    [DataMember]
    private int ColorOfLogo;
}
```

在下面的代码中演示了编写 CompanyLogo 类型的正确方法。

```
[DataContract]
[KnownType(typeof(CircleType))]
[KnownType(typeof(TriangleType))]
public class CompanyLogo2
{
    [DataMember]
    private Shape ShapeOfLogo;
    [DataMember]
    private int ColorOfLogo;
```

只要反序列化外部类型 CompanyLogo2，反序列化引擎就会了解有关 CircleType 和 TriangleType，因此能够查找"Circle"和"Triangle"数据协定的匹配类型。

[示例 2]

在下面的示例中，尽管 CustomerTypeA 和 CustomerTypeB 都具有 Customer 数据协定，但是只要反序列化 CustomerTypeB 就会创建 PurchaseOrder 的实例，因为只有 CustomerTypeB 对反序列化引擎是已知的。

```
public interface ICustomerInfo
{
    string ReturnCustomerName();
}

[DataContract(Name = "Customer")]
public class CustomerTypeA : ICustomerInfo
{
    public string ReturnCustomerName()
    {
        return "no name";
    }
}

[DataContract(Name = "Customer")]
public class CustomerTypeB : ICustomerInfo
{
    public string ReturnCustomerName()
    {
        return "no name";
    }
}

[DataContract]
[KnownType(typeof(CustomerTypeB))]
public class PurchaseOrder
{
    [DataMember]
    ICustomerInfo buyer;
```

```
[DataMember]
int amount;
}
```

[示例 3]

在下面的示例中，Hashtable 将其内容在内部存储为 Object。若要成功反序列化哈希表，反序列化引擎必须知道那里可能出现的一组可能类型。在这种情况下，我们预先知道只有 Book 和 Magazine 对象存储在 Catalog 中，因此使用 KnownTypeAttribute 属性添加它们。

```
[DataContract]
public class Book { }

[DataContract]
public class Magazine { }

[DataContract]
[KnownType(typeof(Book))]
[KnownType(typeof(Magazine))]
public class LibraryCatalog
{
    [DataMember]
    System.Collections.Hashtable theCatalog;
}
```

[示例 4]

在下面的示例中，数据协定存储一个数字和要对该数字执行的操作。Numbers 数据成员可以是整数、整数数组或包含整数的 List < T >。

小心

仅当使用 SVCUTIL.EXE 来生成 WCF 代理时，此方法才能在客户端使用。SVCUTIL.EXE 从包含任何已知类型的服务中检索元数据。如果没有此信息，客户端将不能反序列化该类型。

```
[DataContract]
[KnownType(typeof(int[]))]
public class MathOperationData
{
    private object numberValue;
    [DataMember]
```

```csharp
        public object Numbers
        {
            get { return numberValue; }
            set { numberValue = value; }
        }
        //[DataMember]
        //public Operation Operation;
}
```

这是应用程序代码。

```csharp
//This is in the service application code:
static void Run()
{

    MathOperationData md = new MathOperationData();

    //This will serialize and deserialize successfully because primitive
    //types like int are always known.
    int a = 100;
    md.Numbers = a;

    //This will serialize and deserialize successfully because the array of
    //integers was added to known types.
    int[] b = new int[100];
    md.Numbers = b;

    //This will serialize and deserialize successfully because the generic
    //List<int> is equivalent to int[], which was added to known types.
    List<int> c = new List<int>();
    md.Numbers = c;
    //This will serialize but will not deserialize successfully because
    //ArrayList is a non-generic collection, which is equivalent to
    //an array of type object. To make it succeed, object[]
    //must be added to the known types.
    ArrayList d = new ArrayList();
    md.Numbers = d;

}
```

4. 已知类型、继承和接口

使用 KnownTypeAttribute 属性将已知类型与特定类型关联时,已知类型也与该类型的所有派生类型关联。例如,请参见下面的代码。

```
[DataContract]
[KnownType(typeof(Square))]
[KnownType(typeof(Circle))]
public class MyDrawing
{
    [DataMember]
    private object Shape;
    [DataMember]
    private int Color;
}


[DataContract]
public class DoubleDrawing : MyDrawing
{
    [DataMember]
    private object additionalShape;
}
```

DoubleDrawing 类无需 KnownTypeAttribute 属性即可在 Square 字段中使用 Circle 和 AdditionalShape,因为基类(Drawing)已经应用这些属性。

已知类型只能与类和结构关联,而不能与接口关联。

5. 使用开放式泛型方法的已知类型

可能需要将泛型类型作为已知类型添加。但是,不能将开放式泛型类型作为参数传递到 KnownTypeAttribute 属性。

通过使用替代机制可以解决此问题:编写一个返回要添加到已知类型集合的类型列表的方法。然后将方法名称指定为 KnownTypeAttribute 属性的字符串参数(由于某些限制所致)。

方法必须存在于应用 KnownTypeAttribute 属性的类型上,不得接受参数,且必须返回可以分配给 IEnumerable 的 Type 的对象。

不能将具有方法名称的 KnownTypeAttribute 属性与实际类型在同一类型上的 KnownTypeAttribute 属性组合在一起。此外,不能将具有方法名称的多个 KnownTypeAttribute 应用于同一类型。

请参见下面的类。
```
[DataContract]
public class DrawingRecord<T>
{
    [DataMember]
    private T theData;
    [DataMember]
    private GenericDrawing<T> theDrawing;
}
```

theDrawing 字段包含泛型类 ColorDrawing 和泛型类 BlackAndWhiteDrawing 的实例,这两个泛型类都是从泛型类 Drawing 继承的。通常,必须将它们添加到已知类型,但下面不是有效的属性语法。

```
// Invalid syntax for attributes:
//[KnownType(typeof(ColorDrawing<T>))]
//[KnownType(typeof(BlackAndWhiteDrawing<T>))]
```

因此,必须创建一个方法以返回这些类型。在下面的代码中演示了编写此类型的正确方法。

```
[DataContract]
[KnownType("GetKnownType")]
public class DrawingRecord2<T>
{
    [DataMember]
    private T TheData;
    [DataMember]
    private GenericDrawing<T> TheDrawing;

    private static Type[] GetKnownType()
    {
        Type[] t = new Type[2];
        t[0] = typeof(ColorDrawing<T>);
        t[1] = typeof(BlackAndWhiteDrawing<T>);
        return t;
    }
}
```

6. 添加已知类型的其他方法

此外,可以通过配置文件添加已知类型。在不控制需要已知类型才能正确反序列化的类型时,这是很有用的,如将第三方类型库与 Windows Communication Foundation（WCF）一起使用时。

下面的配置文件演示如何在配置文件中指定已知类型。

```
<configuration>
<system.runtime.serialization>
<dataContractSerializer>
<declaredTypes>
<add type="MyCompany.Library.Shape,
MyAssembly, Version=2.0.0.0, Culture=neutral,
PublicKeyToken=XXXXXX, processorArchitecture=MSIL">
<knownType type="MyCompany.Library.Circle,
MyAssembly, Version=2.0.0.0, Culture=neutral,
PublicKeyToken=XXXXXX, processorArchitecture=MSIL"/>
</add>
</declaredTypes>
</dataContractSerializer>
</system.runtime.serialization>
</configuration>
```

在前面的配置文件中,名为 MyCompany.Library.Shape 的数据协定类型被声明具有已知类型 MyCompany.Library.Circle。

3.5 数据协定序列化程序

Windows Communication Foundation（WCF）基础结构使用 DataContractSerializer 类作为默认序列化程序。

3.5.1 序列化和反序列化

Windows Communication Foundation（WCF）包括新序列化引擎 DataContractSerializer。DataContractSerializer 可在 .NET Framework 对象和 XML 之间进行双向转换。本主题说明序列化程序的工作原理。

在对 .NET Framework 对象进行序列化时,序列化程序了解各种序列化编程模型,包括新的数据协定模型。

当对 XML 进行反序列化时,序列化程序使用 XmlReader 和 XmlWriter 类。在某些情况下（例如在使用 XmlDictionaryReader 二进制 XML 格式时),序列化程序也支持

XmlDictionaryWriter 和 WCF 类以使其能够生成优化的 XML。

WCF 还包括一个伴随序列化程序 NetDataContractSerializer。NetDataContractSerializer 与 BinaryFormatter 和 SoapFormatter 序列化程序类似,因为它也发出 .NET Framework 类型名称作为序列化数据的一部分。当在序列化和反序列化结束阶段共享相同的类型时使用此序列化程序。DataContractSerializer 和 NetDataContractSerializer 都派生自公共基类 XmlObjectSerializer。

警告

DataContractSerializer 将包含带小于 20 的十六进制值的控制字符序列化为 XML 实体。将此类数据发送到 WCF 服务时,可能会导致非 WCF 客户端问题。

1. 创建 DataContractSerializer 实例

构造 DataContractSerializer 的实例是一个重要步骤。完成构造后,将不能够更改任何设置。

指定根类型:

根类型是序列化或反序列化实例的类型。DataContractSerializer 有许多构造函数重载,但必须使用 type 参数提供至少一个根类型。

为某个根类型创建的序列化程序不能用于序列化(或反序列化)其他类型,除非该类型是从根类型派生的。下面的示例演示了两个类。

```
[DataContract]
public class Person
{
    //Code not shown.
}
[DataContract]
public class PurchaseOrder
{
    //Code not shown.
}
```

此代码构造 DataContractSerializer 的一个实例,它仅可用于序列化或反序列化 Person 类的实例。

```
DataContractSerializer dcs = new DataContractSerializer(typeof(Person));
//This can now be used to serialize/deserialize Person but not PurchaseOrder.
```

2. 指定已知类型

如果在进行序列化的类型中涉及多态性,并且尚未使用 KnownTypeAttribute 特性或某种其他机制进行处理,则必须使用 knownTypes 参数将可能的已知类型的列表传递给序列化程序的构造函数。

下面的示例演示 LibraryPatron 类,该类包含特定类型 LibraryItem 的集合。第二个类定义 LibraryItem 类型。第三个和第四个类(Book 和 Newspaper)从 LibraryItem 类继承。

```
[DataContract]
public class LibraryPatron
{
    [DataMember]
    public LibraryItem[] borrowedItems;
}
[DataContract]
public class LibraryItem
{
    //Code not shown.
}

[DataContract]
public class Book : LibraryItem
{
    //Code not shown.
}

[DataContract]
public class Newspaper : LibraryItem
{
    //Code not shown.
}
```

下面的代码构造一个使用 knownTypes 参数的序列化程序的实例。

```
//Create a serializer for the inherited types using the knownType parameter.
Type[] knownTypes = new Type[] { typeof(Book), typeof(Newspaper) };
DataContractSerializer dcs =
new DataContractSerializer(typeof(LibraryPatron), knownTypes);
//All types are known after construction.
```

3. 指定默认根名称和命名空间

通常,在对对象进行序列化时,将根据数据协定名称和命名空间确定最外面的 XML 元素的默认名称和命名空间。所有内部元素的名称将根据数据成员名称来确定,这些元素的命名空间是数据协定的命名空间。下面的示例设置 Name 和 Namespace 类的构造函数中的

DataContractAttribute 和 DataMemberAttribute 值。

```
[DataContract(Name = "PersonContract", Namespace = "http://schemas.contoso.com")]
public class Person2
{
    [DataMember(Name = "AddressMember")]
    public Address theAddress;
}

[DataContract(Name = "AddressContract", Namespace = "http://schemas.contoso.com")]
public class Address
{
    [DataMember(Name = "StreetMember")]
    public string street;
}
```

对 Person 类的实例进行序列化将生成类似如下的 XML：

```
<PersonContract xmlns = "http://schemas.contoso.com">
  <AddressMember>
    <StreetMember>123 Main Street</StreetMember>
  </AddressMember>
</PersonContract>
```

但是，可以通过将 rootName 和 rootNamespace 参数的值传递到 DataContractSerializer 构造函数，自定义根元素的默认名称和命名空间。注意，rootNamespace 不会影响对应于数据成员的所包含元素的命名空间，而只是影响最外面元素的命名空间。

可以作为字符串或 XmlDictionaryString 类的实例来传递这些值，从而允许使用二进制 XML 格式对其进行优化。

4. 设置最大对象配额

一些 DataContractSerializer 构造函数重载具有 maxItemsInObjectGraph 参数。此参数确定序列化程序在单个 ReadObject 方法调用中序列化或反序列化的对象的最大数目。（该方法总是读取一个根对象，但此对象的数据成员中可以具有其他对象。这些对象又可以具有其他对象，依此类推。）默认值为 65536。请注意，当序列化或反序列化数组时，每个数组项都计为一个单独的对象。此外还应注意，一些对象可以有大内存表示形式，因此，单独使用此配额可能不足以防范拒绝服务攻击。如果需要增加此配额以至于超出默认值，则一定要在发送（序列化）和接收（反序列化）方同时增加此配额，原因是在读取和写入数据时此配额会同时应用于发送方和接收方。

5. 往返行程

在一次操作中对对象进行反序列化和重新序列化时将发生往返行程。因此,往返行程是从 XML 到对象实例,然后再返回到 XML 流。

一些 DataContractSerializer 构造函数重载具有 ignoreExtensionDataObject 参数,该参数默认设置为 false。在此默认模式中,对于一个往返行程,可以将数据从数据协定的较新版本发送到较旧版本然后再返回到较新版本而不会出现任何损失,前提是数据协定实现 IExtensibleDataObject 接口。例如,假设 Person 数据协定的版本 1 包含 Name 和 PhoneNumber 数据成员,并且版本 2 添加 Nickname 成员。如果在从版本 2 向版本 1 发送信息时实现了 IExtensibleDataObject,则会存储 Nickname 数据,并在再次序列化数据时重新发出这些数据,因此,在往返行程中不会出现数据丢失。有关详细信息,请参见向前兼容的数据协定和数据协定版本管理。

往返行程的安全性和架构有效性问题:

往返行程可能会涉及一些安全性问题。例如,反序列化和存储大量外来数据可能存在安全风险。重新发出无法验证的数据可能会存在安全问题,尤其是在涉及数字签名的情况下。例如,在前面的方案中,版本 1 终结点可能会对包含恶意数据的 Nickname 值进行签名。最后,还可能存在架构有效性问题:终结点可能需要始终发出严格符合其声明的协定并且没有任何额外值的数据。在前面的示例中,版本 1 终结点的协定声明该终结点仅发出 Name 和 PhoneNumber,并且如果正在使用构造验证,则发出额外的 Nickname 值将导致验证失败。

6. 启用和禁用往返行程

要关闭往返行程,请不要实现 IExtensibleDataObject 接口。如果用户无法控制相应的类型,则将 ignoreExtensionDataObject 参数设置为 true 也可获得同样的效果。

通常,序列化程序并不关心对象标识,如下面的代码中所示:

```
[DataContract]
public class PurchaseOrder
{
    [DataMember]
    public Address billTo;
    [DataMember]
    public Address shipTo;
}

[DataContract]
public class Address
```

```
{
    [DataMember]
    public string street;
}
```

下面的代码创建一份订单:

```
//Construct a purchase order:
Address adr = new Address();
adr.street = "123 Main St.";
PurchaseOrder po = new PurchaseOrder();
po.billTo = adr;
po.shipTo = adr;
```

请注意,将 billTo 和 shipTo 字段设置为同一个对象实例。但是,生成的 XML 会复制重复的信息,并与下面的 XML 类似。

```
<PurchaseOrder>
  <billTo><street>123 Main St.</street></billTo>
  <shipTo><street>123 Main St.</street></shipTo>
</PurchaseOrder>
```

不过,此方法具有以下可能不需要的特征:

◇ 性能。复制数据的效率低。
◇ 循环引用。如果对象引用自身,甚至通过其他对象引用自身,则通过复制进行序列化会导致无限循环。(如果发生这种状况,序列化程序将引发 SerializationException)
◇ 语义。有时,一定要记住这一点:两个引用指向的是同一个对象而不是两个相同的对象。

有关这些原因,一些 DataContractSerializer 构造函数重载具有 preserveObjectReferences 参数(默认值为 false)。在将此参数设置为 true 时,将使用只有 WCF 才可以理解的编码引用的特殊方法。当设置为 true 时,XML 代码示例如下所示:

```
<PurchaseOrder ser:id="1">
  <billTo ser:id="2"><street ser:id="3">123 Main St.</street></billTo>
  <shipTo ser:ref="2"/>
</PurchaseOrder>
```

"ser"命名空间引用标准序列化命名空间 http://schemas.microsoft.com/2003/10/Serialization/。每一段数据只进行一次序列化并获得一个 ID 号,后续使用会导致引用已序列化的数据。

重要

如果"id"和"ref"属性同时存在于数据协定 XMLElement 中,则接受"ref"属性,而忽略

"id"属性。

了解此模式的限制是很重要的:

DataContractSerializer 在 preserveObjectReferences 设置为 true 的情况下生成的 XML 与任何其他技术都无法进行交互,仅可以由另一个其 DataContractSerializer 也设置为 preserveObjectReferences 的 true 实例进行访问。

元数据(架构)不支持此功能。生成的架构仅对 preserveObjectReferences 设置为 false 的情况有效。

此功能可能导致序列化和反序列化进程运行速度减慢。尽管不必复制数据,但是在此模式中必须执行额外的对象比较。

小心

当启用 preserveObjectReferences 模式时,需要将 maxItemsInObjectGraph 值设置为正确的配额,这一点特别重要。由于在此模式中处理数组的方式方面的原因,攻击者很容易构造一条小的恶意消息来造成内存大量消耗(仅通过 maxItemsInObjectGraph 配额来限制)。

7. 指定数据协定代理项

一些 DataContractSerializer 构造函数重载具有 dataContractSurrogate 参数,该参数可以设置为 null。此外,可以使用它来指定数据协定代理项,数据协定代理项是一种实现 IDataContractSurrogate 接口的类型。然后可以使用该接口来自定义序列化和反序列化进程。

8. 序列化

下面的信息适用于从 XmlObjectSerializer 继承的任何类,包括 DataContractSerializer 和 NetDataContractSerializer 类。

(1)简单序列化

对对象进行序列化最基本的方法是将其传递到 WriteObject 方法。该方法有三个重载,每个重载分别用于写入 Stream、XmlWriter 或 XmlDictionaryWriter。使用 Stream 重载时,输出是采用 UTF-8 编码的 XML。使用 XmlDictionaryWriter 重载时,序列化程序会针对二进制 XML 优化其输出。

使用 WriteObject 方法时,序列化程序包装元素的默认名称和命名空间,并将其内容一起写出。

下面的示例演示如何使用 XmlDictionaryWriter 进行写入。

```
Person p = new Person();
DataContractSerializer dcs =
    new DataContractSerializer(typeof(Person));
XmlDictionaryWriter xdw =
    XmlDictionaryWriter.CreateTextWriter(someStream,Encoding.UTF8);
```

```
dcs.WriteObject(xdw, p);
```
这将生成类似于如下所示的 XML：
```
<Person>
    <Name>Jay Hamlin</Name>
    <Address>123 Main St.</Address>
</Person>
```

（2）分步引导的序列化

WriteStartObject、WriteObjectContent 和 WriteEndObject 方法可分别用于写入结束元素、写入对象内容以及关闭包装元素。

备注

这些方法没有 Stream 重载。

此分步引导的序列化具有两个常见用途。一种用途是在 WriteStartObject 和 WriteObjectContent 之间插入内容（例如属性或注释），如以下示例所示。

```
dcs.WriteStartObject(xdw, p);
xdw.WriteAttributeString("serializedBy", "myCode");
dcs.WriteObjectContent(xdw, p);
dcs.WriteEndObject(xdw);
```

这将生成类似于如下所示的 XML：
```
<Person serializedBy="myCode">
    <Name>Jay Hamlin</Name>
    <Address>123 Main St.</Address>
</Person>
```

另一种常见用途是完全避免使用 WriteStartObject 和 WriteEndObject，并写入自己的自定义包装元素（或者甚至连同跳过写入包装），如以下代码中所示。

```
xdw.WriteStartElement("MyCustomWrapper");
dcs.WriteObjectContent(xdw, p);
xdw.WriteEndElement();
```

这将生成类似于如下所示的 XML：
```
<MyCustomWrapper>
    <Name>Jay Hamlin</Name>
    <Address>123 Main St.</Address>
</MyCustomWrapper>
```

备注

使用分步引导的序列化可能会导致架构无效的 XML。

9. 反序列化

下面的信息适用于从 XmlObjectSerializer 继承的任何类,包括 DataContractSerializer 和 NetDataContractSerializer 类。

对对象进行反序列化的最基本的方式是调用 ReadObject 方法重载之一。该方法有三个重载,每个重载分别用于读取 XmlDictionaryReader、XmlReader 或 Stream。请注意,Stream 重载将创建不受任何配额保护的文本 XmlDictionaryReader,此重载仅应用于读取受信任的数据。

还请注意,必须将 ReadObject 方法返回的对象强制转换为适当的类型。

下面的代码构造 DataContractSerializer 和 XmlDictionaryReader 的实例,然后对 Person 实例进行反序列化。

```
DataContractSerializer dcs = new DataContractSerializer(typeof(Person));
FileStream fs = new FileStream(path, FileMode.Open);
XmlDictionaryReader reader =
XmlDictionaryReader.CreateTextReader(fs, new XmlDictionaryReaderQuotas());
Person p = (Person)dcs.ReadObject(reader);
```

在调用 ReadObject 方法之前,将 XML 读取器置于包装元素上或包装元素前面的非内容节点上。可以通过调用 Read 或其派生项的 XmlReader 方法并测试 NodeType 来完成此操作,如以下代码所示:

```
DataContractSerializer ser = new DataContractSerializer(typeof(Person),
"Customer", @"http://www.contoso.com");
FileStream fs = new FileStream(path, FileMode.Open);
XmlDictionaryReader reader =
XmlDictionaryReader.CreateTextReader(fs, new XmlDictionaryReaderQuotas());
while (reader.Read())
{
    switch (reader.NodeType)
    {
        case XmlNodeType.Element:
            if (ser.IsStartObject(reader))
            {
                Console.WriteLine("Found the element");
                Person p = (Person)ser.ReadObject(reader);
                Console.WriteLine("{0} {1}    id:{2}",
                    p.Name , p.Address);
            }
```

```
            Console.WriteLine(reader.Name);
            break;
    }
}
```

请注意,在将读取器传递给 ReadObject 之前,可以读取此包装元素上的属性。

当使用一个简单 ReadObject 重载时,反序列化程序会查找默认名称和包装元素上的命名空间(请参见上一部分中,"指定默认根名称和命名空间")和发现为未知时引发异常元素。在上面的示例中,应有 <Person> 包装元素。可调用 IsStartObject 方法来验证是否已将读取器定位在按预期命名的元素上。

有一种方法可以用来禁用此包装元素名称检查;一些 ReadObject 方法的重载采用布尔参数 verifyObjectName,该参数默认设置为 true。当该参数设置为 false 时,包装元素的名称和命名空间将被忽略。这对于读取使用分步引导的序列化机制写入的 XML 是有用的。

10. 使用 NetDataContractSerializer

DataContractSerializer 和 NetDataContractSerializer 之间的主要差异在于 DataContractSerializer 使用数据协定名称,而 NetDataContractSerializer 在序列化的 XML 中输出完整的 .NET Framework 程序集和类型名称。这意味着必须在序列化终结点和反序列化终结点之间共享完全相同的类型。这也同时意味着不需要对 NetDataContractSerializer 使用已知类型机制,因为要反序列化的确切类型始终是已知的。

但是,会出现以下几个问题:

- ◇ 安全性。在要反序列化的 XML 中找到的任何类型都会加载。有人可以利用这一点来强制加载恶意类型。仅在使用 NetDataContractSerializer 序列化联编程序时(使用属性或构造函数参数)才应将 Binder 用于不受信任的数据。联编程序仅允许加载安全类型。联编程序机制与 System.Runtime.Serialization 命名空间中的类型使用的机制相同。

- ◇ 版本管理。在 XML 中使用完整的类型和程序集名称会严格限制对类型进行版本管理的方式。以下内容不可更改:类型名称、命名空间、程序集名称和程序集版本。通过将 AssemblyFormat 属性或构造函数参数设置为 Simple(而非 Full 的默认值),可以允许程序集版本更改,但不允许泛型参数类型更改。

- ◇ 互操作性。由于 .NET Framework 类型和程序集名称包含在 XML 中,因此 .NET Framework 以外的平台不能访问生成的数据。

- ◇ 性能。写出类型和程序集名称会显著增加生成的 XML 的大小。

此机制与 .NET Framework 远程处理(具体是指 BinaryFormatter 和 SoapFormatter)使用的二进制或 SOAP 序列化类似。

使用 NetDataContractSerializer 与使用 DataContractSerializer 类似,但存在以下区别:

◇ 构造函数不要求指定根类型。可以使用 NetDataContractSerializer 的相同实例对任何类型进行序列化。

◇ 构造函数不接受已知类型的列表。如果将类型名称序列化为 XML,则不需要已知类型机制。

◇ 构造函数不接受数据协定代理项,而是接受一个名为 ISurrogateSelector 的 surrogateSelector 参数(映射到 SurrogateSelector 属性)。这是旧式代理项机制。

◇ 构造函数接受 assemblyFormat 的一个名为 FormatterAssemblyStyle 的参数,该参数映射到 AssemblyFormat 属性。如前所述,这可以用于增强序列化程序的版本管理功能。这与二进制或 SOAP 序列化中的 FormatterAssemblyStyle 机制相同。

◇ 构造函数接受一个名为 StreamingContext 的 context 参数,该参数映射到 Context 属性。可以使用该参数将信息传递到要序列化的类型中。此用法与其他 StreamingContext 类中使用的 System.Runtime.Serialization 机制的用法相同。

◇ Serialize 和 Deserialize 方法是 WriteObject 和 ReadObject 方法的别名。这些别名可以为二进制或 SOAP 序列化提供更为一致的编程模型。

NetDataContractSerializer 和 DataContractSerializer 使用的 XML 格式通常是不兼容的。也就是说,不支持尝试使用这些序列化程序的一种进行序列化而使用另一种序列化程序进行反序列化的情况。

还请注意,NetDataContractSerializer 对于对象图中的每个节点不会输出完整的 .NET Framework 类型和程序集名称。仅在有歧义的地方才会输出上述信息。也就是说,它是在根对象级别进行输出并且是针对任何多态情况。

3.5.2 使用数据协定解析程序

使用数据协定解析程序可以动态配置已知类型。序列化或反序列化并非数据协定所需的类型时,要求提供已知类型。已知类型通常以静态方式指定。这意味着用户必须了解在实现某个操作期间,该操作可能接收的所有可能类型。在某些方案中无法做到这一点,因此能够以动态方式指定已知类型十分重要。

1. 创建数据协定解析程序

创建数据协定解析程序实现有两个方法:TryResolveType 和 ResolveName。这两个方法分别实现在序列化和反序列化期间使用的回调。在序列化期间将调用 TryResolveType 方法,用于获取数据协定类型并将其映射到 xsi:type 名称和命名空间。在反序列化期间将调用 ResolveName 方法,用于获取 xsi:type 名称和命名空间并将其解析为数据协定类型。这两个方法均具有 knownTypeResolver 参数,该参数可用于在实现中使用默认已知类型解析程序。

下面的示例演示了如何实现 DataContractResolver，以映射到派生自数据协定类型 Customer 的数据协定类型 Person，或者从后一个数据协定类型进行映射。

```
public class MyCustomerResolver : DataContractResolver
{
    public override bool TryResolveType ( Type dataContractType, Type declaredType, DataContractResolver knownTypeResolver, out XmlDictionaryString typeName, out XmlDictionaryString typeNamespace)
    {
        if (dataContractType = = typeof(Customer))
        {
            XmlDictionary dictionary = new XmlDictionary();
            typeName = dictionary.Add("SomeCustomer");
            typeNamespace = dictionary.Add("http://tempuri.com");
            return true;
        }
        else
        {
            return knownTypeResolver.TryResolveType(dataContractType, declaredType, null, out typeName, out typeNamespace);
        }
    }

    public override Type ResolveName ( string typeName, string typeNamespace, DataContractResolver knownTypeResolver)
    {
        if (typeName = = "SomeCustomer" && typeNamespace = = "http://tempuri.com")
        {
            return typeof(Customer);
        }
        else
        {
            return knownTypeResolver.ResolveName(typeName, typeNamespace, null);
        }
    }
}
```

一旦定义了 DataContractResolver，即可将它传递到 DataContractSerializer 构造函数加以

第3章 数据协定

使用，如下面的示例所示。
```
XmlObjectSerializer serializer = new DataContractSerializer (typeof (Customer), null,
Int32.MaxValue, false, false, null, new MyCustomerResolver());
```
可以在对 DataContractSerializer 或 ReadObject 方法的调用中指定 WriteObject，如下面的示例所示。
```
MemoryStream ms = new MemoryStream();
DataContractSerializer serializer = new DataContractSerializer (typeof
(Customer));
XmlDictionaryWriter writer = XmlDictionaryWriter.CreateDictionaryWriter
(XmlWriter.Create(ms));
serializer.WriteObject(writer, new Customer(), new MyCustomerResolver());
writer.Flush();
ms.Position = 0;
Console.WriteLine (((Customer) serializer.ReadObject (XmlDictionaryReader.
CreateDictionaryReader (XmlReader.Create (ms)), false, new MyCustomerResolver
()));
```
或者，可以在 DataContractSerializerOperationBehavior 上设置该构造函数，如下面的示例所示。
```
ServiceHost host = new ServiceHost(typeof(MyService));

ContractDescription cd = host.Description.Endpoints[0].Contract;
OperationDescription myOperationDescription = cd.Operations.Find("Echo");

DataContractSerializerOperationBehavior serializerBehavior =
myOperationDescription.Behaviors.Find<DataContractSerializerOperationBehavior
>();
if (serializerBehavior == null)
{
    serializerBehavior = new DataContractSerializerOperationBehavior
(myOperationDescription);
    myOperationDescription.Behaviors.Add(serializerBehavior);
}
SerializerBehavior.DataContractResolver = new MyCustomerResolver();
```
通过实现可以应用于服务的特性，可以通过声明方式指定数据协定解析程序。有关详细信息，请参见 KnownAssemblyAttribute 示例。此示例实现一个称为"KnownAssembly"的属性，它将自定义数据协定解析程序添加到服务的行为。

第 4 章 事 务

4.1 事务概述

事务可提供一种分组方法，将一组操作分为单个不可分的执行单元。事务是指具有下列属性的操作集合：
- ◇ 原子性。此属性可确保特定事务下完成的所有更新都已提交并保持持久，或所有这些更新都已中止并回滚到其先前状态。
- ◇ 一致性。此属性可保证某一事务下所做的更改表示从一种一致状态转换到另一种一致状态。例如，将钱从支票账户转移到存款账户的事务并不改变整个银行账户中的钱的总额。
- ◇ 隔离。此属性可防止事务遵循属于其他并发事务的未提交的更改。隔离在确保一种事务不能对另一事务的执行产生意外的影响的同时，还提供一个抽象的并发。
- ◇ 持续性。这意味着一旦提交对托管资源（如数据库记录）的更新，即使出现失败，这些更新也会保持持久。

Windows Communication Foundation（WCF）提供一组丰富的功能，使用户能够在 Web 服务应用程序中创建分布式事务。

WCF 实现对 WS–AtomicTransaction（WS–AT）协议的支持，该协议使 WCF 应用程序能够将事务传输到可互操作应用程序中，例如使用第三方技术生成的可互操作 Web 服务。WCF 还实现对 OLE 事务协议的支持，可以在无需互操作功能以实现事务流式处理的情况下使用该协议。

用户可以使用应用程序配置文件来配置绑定以启用或禁用事务流，以及设置有关绑定的所需事务协定。此外，用户还可以使用配置文件在服务级别设置事务超时值。

System.ServiceModel 命名空间中的事务属性允许用户进行以下操作：
- ◇ 使用 ServiceBehaviorAttribute 属性配置事务超时值以及隔离级别的筛选。
- ◇ 启用事务功能并使用 OperationBehaviorAttribute 属性配置事务完成行为。
- ◇ 使用协定方法上的 ServiceContractAttribute 和 OperationContractAttribute 属性来要求、允许或拒绝事务流。

4.2 事务性支持

4.2.1 ServiceModel 事务特性

Windows Communication Foundation（WCF）在以下三个标准 System.ServiceModel 属性（Attribute）上提供用于配置 WCF 服务的事务行为的属性（Property）：

- ◇ TransactionFlowAttribute；
- ◇ ServiceBehaviorAttribute；
- ◇ OperationBehaviorAttribute。

1. TransactionFlowAttribute

TransactionFlowAttribute 属性指定服务协定中的操作是否愿意接受来自客户端的传入事务。此属性（Attribute）通过以下属性（Property）提供此控制：事务使用 TransactionFlowOption 枚举指定传入事务是 Mandatory、Allowed 还是 NotAllowed。

此属性是将服务操作与客户端的外部交互操作相关联的唯一属性。下面几节中说明的属性与在操作的执行中使用的事务有关。

2. ServiceBehaviorAttribute

ServiceBehaviorAttribute 属性指定服务协定实现的内部执行行为。此属性（Attribute）特定于事务的属性（Property）包括：

- ◇ TransactionAutoCompleteOnSessionClose，此属性指定会话关闭时是否完成未完成的事务。此属性的默认值为 false。如果此属性为 true 且传入会话正常关闭而不是由于网络或客户端故障而关闭，则会成功完成任何未完成的事务。否则，如果此属性为 false 或者如果会话未正常关闭，则会话关闭时任何未完成的事务将会回滚。如果此属性为 true，则传入通道必须基于会话。
- ◇ ReleaseServiceInstanceOnTransactionComplete，此属性指定事务完成时是否释放基础服务实例。此属性的默认值为 true。下一个入站消息会导致创建新的基础实例，放弃上一个实例可能保持的每个事务的任何状态。释放服务实例是服务执行的内部操作，对客户端可能已经建立的任何现有连接或会话没有影响。此功能等效于 COM+ 提供的实时激活功能。如果此属性为 true，则 ConcurrencyMode 必须等于 Single。否则，服务在启动过程中会引发无效配置验证异常。
- ◇ TransactionIsolationLevel，此属性指定用于服务内事务的隔离级别；此属性采用 IsolationLevel 值之一。如果本地隔离级别属性是 Unspecified 以外的任何值，则传入事务的隔离级别必须与此本地属性的设置相匹配，否则会拒绝传入事务并将故障发回客户端。如果 TransactionScopeRequired 为 true，且没有对事务进行流处理，则此属性确定要用于本地创建的事务的 IsolationLevel 值。如果 IsolationLevel 设置为 Unspecified，IsolationLevel Serializable 使用。

◇ TransactionTimeout，此属性指定一个时间段，在服务中创建的新事务必须在此时间段内完成。如果达到此时间时事务没有完成，则会中止事务。对于已将 TimeSpan 设置为 TransactionScope 的任何操作以及为其创建了新事务的任何操作，TransactionScopeRequired 用作 true 超时。该超时是从创建事务到完成两阶段提交协议的第 1 阶段所允许的最长时间。使用的超时值始终是 TransactionTimeout 属性和 transactionTimeout 配置设置之间的较小值。

3. OperationBehaviorAttribute

OperationBehaviorAttribute 属性指定服务实现中方法的行为。可以用此属性指示操作的特定执行行为。此属性（Attribute）的属性（Property）不影响服务协定的 Web 服务描述语言（WSDL）说明，它们只是用于启用通用功能的 WCF 编程模型元素，如果没有这些属性（Property），开发人员将不得不自己实现这些功能。

此属性（Attribute）具有以下特定于事务的属性（Property）：

◇ TransactionScopeRequired 指定是否必须在活动事务范围内执行方法。默认值为 false。如果没有为方法设置 OperationBehaviorAttribute 属性，这也暗示着不会在事务中执行该方法。如果操作不需要事务范围，则不会激活消息头中存在的任何事务，并且这些事务将保持作为 IncomingMessageProperties 的 OperationContext 的元素。如果操作需要事务范围，则从下列各项之一中派生该事务的源：
  √ 如果从客户端流动事务，则在使用该分布式事务创建的事务范围内执行此方法。
  √ 使用排队传输时，用于对消息取消排队的事务。请注意，使用的事务不是流事务，因为它不是由消息的原始发送方提供。
  √ 自定义传输可以通过使用 TransportTransactionProperty 提供事务。
  √ 如果上面的任一属性均没有为事务提供外部源，则会在调用方法之前创建一个新的 Transaction 实例。

◇ TransactionAutoComplete 指定在没有引发未处理的异常的情况下，在其中执行方法的事务是否自动完成。如果此属性为 true，则当用户方法在未引发异常的情况下返回时，调用基础结构会自动将事务标记为"已完成"。如果此属性为 false，则事务会附加到实例，并且仅当客户端调用标记为此属性等于 true 的后续方法时，或仅当后续方法显式调 SetTransactionComplete 时，事务才会标记为"已完成"。不执行这两种方案之一会导致事务永远也不会处于"已完成"状态,其中所包含的工作也不会提交,除非将 TransactionAutoCompleteOnSessionClose 属性设置为 true。如果此属性设置为 true,则必须与会话一起使用通道,且必须将 InstanceContextMode 设置为 PerSession。

4.2.2 ServiceModel 事务配置

Windows Communication Foundation（WCF）提供了以下三个用于为服务配置事务的属性:transactionFlow、transactionProtocol 和 transactionTimeout。

1. 配置 transactionFlow

WCF 提供的大多数预定义绑定都包含 transactionFlow 和 transactionProtocol 属性,以便用户可以使用特定的事务流协议为特定终结点配置用于接受传入事务的绑定。此外,用户还可以使用 transactionFlow 元素及其 transactionProtocol 特性生成用户自己的自定义绑定。

transactionFlow 属性指定是否为使用绑定的服务终结点启用事务流。

2. 配置 transactionProtocol

transactionProtocol 属性指定要应用于使用绑定的服务终结点的事务协议。

下面是一个配置节示例,该配置节将指定的绑定配置为支持事务流并且使用 WS – AtomicTransaction 协议。

```
<netNamedPipeBinding>
  <binding name = "test"
    closeTimeout = "00:00:10"
    openTimeout = "00:00:20"
    receiveTimeout = "00:00:30"
    sendTimeout = "00:00:40"
    transactionFlow = "true"
    transactionProtocol = "WSAtomicTransactionOctober2004"
    hostNameComparisonMode = "WeakWildcard"
    maxBufferSize = "1001"
    maxConnections = "123"
    maxReceivedMessageSize = "1000" >
  </binding>
</netNamedPipeBinding>
```

3. 配置 transactionTimeout

可以在配置文件的 transactionTimeout 元素中配置 WCF 服务的 behavior 属性。下面的代码演示如何执行此操作。

```
<configuration>
  <system.serviceModel>
    <behaviors>
      <behavior name = "NewBehavior" transactionTimeout = "00:01:00" /> <!-
- 1 minute timeout - - >
    </behaviors>
  </system.serviceModel>
</configuration>
```

transactionTimeout 属性指定了在该服务中创建的新事务必须在此期间完成的时间段。它被用作任何建立新事务的操作的 TransactionScope 超时, 而且, 如果应用了 OperationBehaviorAttribute, 则 TransactionScopeRequired 属性将设置为 true。

超时指定了从创建事务到完成两阶段提交协议的第 1 阶段之间的持续时间。

如果在 service 配置节内设置了此属性, 则应该通过 OperationBehaviorAttribute 应用相应服务的至少一个方法, 其中 TransactionScopeRequired 属性设置为 true。

请注意, 所使用的超时值是此 transactionTimeout 配置设置和任何 TransactionTimeout 属性之间的较小值。

4.2.3 启用事务流

Windows Communication Foundation (WCF) 提供用于控制事务流的灵活性较高的选项。服务事务流设置可以使用属性与配置的组合来表示。

1. 事务流设置

服务终结点的事务流设置根据下列三个值的交集生成:

◇ 为服务协定中的每个方法指定的 TransactionFlowAttribute 属性。
◇ 特定绑定中的 TransactionFlow 绑定属性。
◇ 特定绑定中的 TransactionFlowProtocol 绑定属性。TransactionFlowProtocol 绑定属性允许用户在可用于流动事务的两个不同事务协议之间进行选择。后面几节将对这些协议逐一进行简要描述。

2. WS – AtomicTransaction 协议

WS – AtomicTransaction (WS – AT) 协议对于要求第三方协议堆栈具有互操作性时的情形非常有用。

3. OleTransactions 协议

OleTransactions 协议对于如下的情形非常有用: 即不要求第三方协议堆栈具有互操作性, 并且服务部署人员预先知道 WS – AT 协议服务将在本地禁用或者现有网络拓扑不支持使用 WS-AT。

表 4-1 列出了可以使用这些不同组合生成的不同类型的事务流。

表 4-1 使用不同组合生成的不同类型的事务流

TransactionFlow 绑定	TransactionFlow 绑定属性	TransactionFlowProtocol 绑定协议	事务流的类型
必需	true	WS-AT	事务必须以可以互操作的 WS-AT 格式流动
强制	true	OleTransactions	事务必须以 WCF OleTransactions 格式流动
强制	False	不适用	不适用,因为这是无效的配置
Allowed	true	WS-AT	事务可以以可互操作的 WS-AT 格式流动
Allowed	true	OleTransactions	事务可以以 WCF OleTransactions 格式流动
Allowed	False	任意值	不流动事务
NotAllowed	任意值	任意值	不流动事务

表 4-2 对消息处理结果做了总结。

表 4-2 消息处理

传入消息	TransactionFlow 设置	事务标头	消息处理结果
事务与预期的协议格式匹配	Allowed 或 Mandatory	MustUnderstand 等于 true	进程
事务不与预期的协议格式匹配	强制	MustUnderstand 等于 false	因要求一个事务而拒绝
事务不与预期的协议格式匹配	Allowed	MustUnderstand 等于 false	因无法理解标头而拒绝
使用任何协议格式的事务	NotAllowed	MustUnderstand 等于 false	因无法理解标头而拒绝
无事务	强制	不可用	因要求一个事务而拒绝
无事务	Allowed	不可用	进程
无事务	NotAllowed	不可用	进程

虽然协定上的每个方法都有不同的事务流要求,但事务流协议设置的范围处于绑定级别。这意味着,共享同一个终结点(并因此共享同一绑定)的所有方法也共享允许或要求事务流的同一个策略以及同一个事务协议(如果适用)。

4. 在方法级别启用事务流

对于服务协定中的所有方法,事务流需求并不总是相同。因此,WCF 还提供基于属性的机制以允许表示每个方法的事务流首选项。这通过指定服务操作接受事务标头所处的级别的 TransactionFlowAttribute 来实现。如果需要启用事务流,则应使用此属性标记服务协定方法。此属性采用 TransactionFlowOption 枚举值之一,其中默认值为 NotAllowed。如果指定除 NotAllowed 以外的任何值,则要求该方法不要成为单向方法。开发人员可以使用此属性在设计时指定方法级别的事务流要求或约束。

5. 在终结点级别启用事务流

除了 TransactionFlowAttribute 属性提供的方法级别的事务流设置以外,WCF 也为事务流提供终结点范围的设置,以允许管理员可以在较高级别控制事务流。

这可以通过 TransactionFlowBindingElement 来实现,该类允许用户在终结点绑定设置中启用或禁用传入事务流,并允许指定传入事务所需的事务协议格式。

如果该绑定已禁用事务流,但对服务协定的操作之一要求一个传入事务,则将在服务启动时引发验证异常。

WCF 提供的大多数持续绑定都包含 transactionFlow 和 transactionProtocol 特性,以允许用户将特定的绑定配置为接受传入事务。

管理员或部署人员可以使用终结点级别的事务流在部署时使用配置文件来配置事务流需求或约束。

6. 安全性

若要确保系统的安全性和完整性,用户必须在应用程序之间流动事务时保护消息交换,不应向无资格参与某一事务的任何应用程序流动或透露该事务的详细信息。

在使用元数据交换向未知或不受信任的 Web 服务生成 WCF 客户端时,对这些 Web 服务上的操作的调用应取消当前的事务(如可能)。下面的示例演示如何执行此操作。

```
//client code which has an ambient transaction
using (TransactionScope scope = new TransactionScope(TransactionScopeOption.Suppress))
{
    //No transaction will flow to this operation
    untrustedProxy.Operation1(...);
    scope.Complete();
}
//remainder of client code
```

此外,还应将这些服务配置为仅接受来自它们已对其进行身份验证和授权的客户端的传入事务。如果传入事务来自高度受信任的客户端,则应仅接受传入事务。

7. 策略断言

WCF 使用策略断言来控制事务流。可以在服务的策略文档中找到策略断言,该断言是通过聚合协定、配置和属性而生成的。客户端可以使用 HTTP GET 或 WS-MetadataExchange 请求-回复来获取服务的策略文档。然后,客户端可以通过处理策略文档来确定服务协定上的哪些操作可以支持或要求事务流。

事务流策略断言通过指定客户端应发送给服务以表示事务的 SOAP 标头来影响事务流。所有事务标头都必须标记 MustUnderstand 等于 true。以其他方式标记标头的任何消息都将被拒绝,并出现 SOAP 错误。

一个操作上只能出现一个与事务相关的策略断言。在一个操作上具有多个事务断言的策略文档将被视为无效,并且将被 WCF 拒绝。此外,每个端口类型内仅可出现一个事务协议。具有引用单个端口类型内的多个事务处理协议的操作的策略文档将被视为无效,并且会拒绝 ServiceModel 元数据实用工具(Svcutil.exe)。事务断言出现在输出消息或单向输入消息上的策略文档也将被视为无效。

4.2.4 创建事务性服务

本示例演示创建事务性服务和使用客户端启动的事务协调服务操作的各个方面。

1. 创建事务性服务

(1)创建服务协定并用 TransactionFlowOption 枚举中所需的设置对操作进行批注以指定传入事务要求。请注意,也可以将 TransactionFlowAttribute 放在要实现的服务类上。这允许接口的单一实现(而不是每个实现)使用这些事务设置。

```
[ServiceContract]
public interface ICalculator
{
    [OperationContract]
    //Use this to require an incoming transaction
    [TransactionFlow(TransactionFlowOption.Mandatory)]
    double Add(double n1, double n2);
    [OperationContract]
    //Use this to permit an incoming transaction
    [TransactionFlow(TransactionFlowOption.Allowed)]
    double Subtract(double n1, double n2);
}
```

(2)创建一个实现类,并使用 ServiceBehaviorAttribute 有选择地指定 TransactionIsolationLevel

和 TransactionTimeout。应该注意的是，默认值为 60 秒的 TransactionTimeout 和默认值为 TransactionIsolationLevel 的 Unspecified 适用于许多情况。对于每个操作，可以根据 OperationBehaviorAttribute 属性的值，使用 TransactionScopeRequired 属性指定在方法中执行的工作是否应在事务范围内发生。在本例中，用于 Add 方法的事务与从客户端流入的强制传入事务相同，用于 Subtract 方法的事务与传入事务相同（如果已从客户端流入了一个事务）或者是一个以隐式方式在本地创建的新事务。

```
[ServiceBehavior(
    TransactionIsolationLevel = System.Transactions.IsolationLevel.Serializable,
    TransactionTimeout = "00:00:45")]
public class CalculatorService : ICalculator
{
    [OperationBehavior(TransactionScopeRequired = true)]
    public double Add(double n1, double n2)
    {
        //Perform transactional operation
        RecordToLog(String.Format("Adding {0} to {1}", n1, n2));
        return n1 + n2;
    }

    [OperationBehavior(TransactionScopeRequired = true)]
    public double Subtract(double n1, double n2)
    {
        //Perform transactional operation
        RecordToLog(String.Format("Subtracting {0} from {1}", n2, n1));
        return n1 - n2;
    }

    private static void RecordToLog(string recordText)
    {
        //Database operations omitted for brevity
        //This is where the transaction provides specific benefit
        // - changes to the database will be committed only when
        //the transaction completes.
    }
}
```

（3）在配置文件中配置绑定，指定事务上下文应进行流处理，并指定要使用的协议执行此操作。有关详细信息，请参见 ServiceModel 事务配置。具体地说，绑定类型是在终结点元素的 binding 属性中指定的。<终结点>元素包含 bindingConfiguration 属性，该属性引用名为 transactionalOleTransactionsTcpBinding 的绑定配置，如下面的示例配置中所示。

```
< service name = "CalculatorService" >
  < endpoint address = "net.tcp://localhost:8008/CalcService"
    binding = "netTcpBinding"
    bindingConfiguration = "transactionalOleTransactionsTcpBinding"
    contract = "ICalculator"
    name = "OleTransactions_endpoint" />
< /service >
```

事务流是在配置级别通过使用 transactionFlow 属性启用的，而事务协议是使用 transactionProtocol 属性指定的，如下面的配置中所示。

```
< bindings >
  < netTcpBinding >
     < binding name = "transactionalOleTransactionsTcpBinding"
       transactionFlow = "true"
       transactionProtocol = "OleTransactions" />
  < /netTcpBinding >
< /bindings >
```

2. 支持多事务协议

为了获得最佳性能，应该对使用 Windows Communication Foundation（WCF）编写的客户端和服务的相关方案使用 OleTransactions 协议。但是，对于需要与第三方协议堆栈之间具有互操作性的方案，可以使用 WS – AtomicTransaction（WS – AT）协议。通过为多个终结点提供特定于协议的恰当绑定，可以将 WCF 服务配置成同时接受这两种协议，如下面的示例配置所示。

```
< service name = "CalculatorService" >
  < endpoint address = "http://localhost:8000/CalcService"
    binding = "wsHttpBinding"
    bindingConfiguration = "transactionalWsatHttpBinding"
    contract = "ICalculator"
    name = "WSAtomicTransaction_endpoint" />
  < endpoint address = "net.tcp://localhost:8008/CalcService"
    binding = "netTcpBinding"
    bindingConfiguration = "transactionalOleTransactionsTcpBinding"
```

```
      contract = "ICalculator"
      name = "OleTransactions_endpoint" />
</service>
```

事务协议是使用 transactionProtocol 属性指定的。但是,系统提供的 wsHttpBinding 中没有此属性,因为此绑定只能使用 WS – AT 协议。

```
<bindings>
  <wsHttpBinding>
    <binding name = "transactionalWsatHttpBinding"
      transactionFlow = "true" />
  </wsHttpBinding>
  <netTcpBinding>
    <binding name = "transactionalOleTransactionsTcpBinding"
      transactionFlow = "true"
      transactionProtocol = "OleTransactions" />
  </netTcpBinding>
</bindings>
```

3. 控制事务的完成

(1) 默认情况下,如果未引发未处理的异常,WCF 操作会自动完成事务。使用 TransactionAutoComplete 属性和 SetTransactionComplete 方法可以修改此行为。当需要某一操作与另一操作(例如借贷操作)在同一个事务中发生时,可以通过将 TransactionAutoComplete 属性设置为 false 来禁用自动完成行为,如下面的 Debit 操作示例所示。在调用 Debit 属性设置为 TransactionAutoComplete 的方法前(如操作 true 所示)或调用 Credit1 方法以将事务显式标记为完成前(如操作 SetTransactionComplete 所示),Credit2 操作使用的事务不会完成。请注意,所示的两个贷记操作仅供演示,更常见的情况是只使用一个贷记操作。

```
[ServiceBehavior]
public class CalculatorService : IAccount
{
    [OperationBehavior(
        TransactionScopeRequired = true, TransactionAutoComplete = false)]
    public void Debit(double n)
    {
        //Perform debit operation

        return;
```

```
    }

    [OperationBehavior(
        TransactionScopeRequired = true, TransactionAutoComplete = true)]
    public void Credit1(double n)
    {
        // Perform credit operation

        return;
    }

    [OperationBehavior(
        TransactionScopeRequired = true, TransactionAutoComplete = false)]
    public void Credit2(double n)
    {
        // Perform alternate credit operation

        OperationContext.Current.SetTransactionComplete();
        return;
    }
}
```

（2）出于事务关联的目的而将 TransactionAutoComplete 属性设置为 false 需要使用会话绑定。这一要求使用 SessionMode 属性在 ServiceContractAttribute 上指定。

```
[ServiceContract(SessionMode = SessionMode.Required)]
public interface IAccount
{
    [OperationContract]
    [TransactionFlow(TransactionFlowOption.Allowed)]
    void Debit(double n);
    [OperationContract]
    [TransactionFlow(TransactionFlowOption.Allowed)]
    void Credit1(double n);
    [OperationContract]
    [TransactionFlow(TransactionFlowOption.Allowed)]
    void Credit2(double n);
}
```

4. 控制事务性服务实例的生存期

WCF 使用 ReleaseServiceInstanceOnTransactionComplete 属性指定事务完成时是否释放基础服务实例。由于此属性值默认为 true(除非配置了其他值),因此 WCF 具有高效且可预见的"实时"激活行为。在后续事务上调用服务可确保新服务实例不会保留前一个事务的状态。虽然此行为通常很有用,但有时可能希望服务实例在事务完成后仍然保持状态。例如,所需的状态或资源句柄成本昂贵,难以检索或重建时就属于这种情况。通过将 ReleaseServiceInstanceOnTransactionComplete 属性设置为 false,可以实现此目的。使用该设置时,实例和任何关联状态将可用于后续调用。使用此设置时,应仔细考虑状态和事务何时清除和完成以及如何清除和完成。下面的示例演示如何通过对 runningTotal 变量保持实例来实现这一目的。

```
[ServiceBehavior(TransactionIsolationLevel = [ServiceBehavior(
    ReleaseServiceInstanceOnTransactionComplete = false)]
public class CalculatorService : ICalculator
{
    double runningTotal = 0;

    [OperationBehavior(TransactionScopeRequired = true)]
    public double Add(double n)
    {
        // Perform transactional operation
        RecordToLog(String.Format("Adding {0} to {1}", n, runningTotal));
        runningTotal = runningTotal + n;
        return runningTotal;
    }

    [OperationBehavior(TransactionScopeRequired = true)]
    public double Subtract(double n)
    {
        // Perform transactional operation
        RecordToLog ( String. Format ( " Subtracting {0} from {1}", n, runningTotal));
        runningTotal = runningTotal - n;
        return runningTotal;
    }

    private static void RecordToLog(string recordText)
```

```
    }
        //Database operations omitted for brevity
    }
}
```

备注

由于实例的生存期是服务的内在行为,并通过 ServiceBehaviorAttribute 属性进行控制,因此不需要修改服务配置或服务协定来设置实例行为。另外,网络上也不包含这种表示形式。

第 5 章 元 数 据

Windows Communication Foundation（WCF）提供了一个基础结构,用于导出、发布、检索和导入服务元数据。WCF 服务使用元数据来描述如何与服务的终结点进行交互,以便工具（如 Svcutil.exe）可以自动生成客户端代码来访问服务。

5.1 元数据体系结构概述

构成 WCF 元数据基础结构的大多数类型驻留在 System.ServiceModel.Description 命名空间中。

WCF 使用 ServiceEndpoint 类描述服务中的终结点。可以使用 WCF 为服务终结点生成元数据或者导入服务元数据以生成 ServiceEndpoint 实例。

WCF 将服务的元数据表示为 MetadataSet 类型的实例,其结构被强附加到在 WS – MetadataExchange 中定义的元数据序列化格式。MetadataSet 类型将实际的服务元数据（如 Web 服务描述语言（WSDL）文档、XML 架构文档或 WS – Policy 表达式）作为 MetadataSection 实例的集合绑定。每个 System.ServiceModel.Description.MetadataSection 实例都包含特定的元数据方言和标识符。System.ServiceModel.Description.MetadataSection 在其 MetadataSection.Metadata 属性中可能包含以下项：

◇ 原始元数据；
◇ 一个 MetadataReference 实例；
◇ 一个 MetadataLocation 实例。

System.ServiceModel.Description.MetadataReference 实例指向另一个元数据交换（MEX）终结点,而 System.ServiceModel.Description.MetadataLocation 实例指向使用 HTTP URL 的元数据文档。WCF 支持使用 WSDL 文件来描述服务终结点、服务协定、绑定、消息交换模式、消息,以及服务实现的故障消息。服务使用的数据类型在 WSDL 文档中使用 XML 架构进行描述。可以使用 WCF 导出和导入服务行为的 WSDL 扩展、协定行为和扩展服务功能的绑定元素。

1. 导出服务元数据

服务并行的标准化表示通过元数据导出服务终结点及将其投影到客户端来实现的过程。若要从 ServiceEndpoint 实例导出元数据,请使用 MetadataExporter 抽象类实现。System. ServiceModel. Description. MetadataExporter 实现生成包装在 MetadataSet 实例中的元数据。

System. ServiceModel. Description. MetadataExporter 类提供了一个框架,用于生成描述终结点绑定的功能和要求及其关联操作、消息和错误的策略表达式。在 PolicyConversionContext 实例中可捕获这些策略表达式。然后 System. ServiceModel. Description. MetadataExporter 实现将这些策略表达式附加到它生成的元数据。

当生成供 System. ServiceModel. Description. MetadataExporter 实现使用的 System. ServiceModel. Channels. BindingElement 对象时,IPolicyExportExtension 调入在 ServiceEndpoint 的绑定中实现 PolicyConversionContext 接口的每个 System. ServiceModel. Description. MetadataExporter。通过在 IPolicyExportExtension 类型的自定义实现上实现 BindingElement 接口,可以导出新策略断言。

WsdlExporter 类型是 System. ServiceModel. Description. MetadataExporter 所包含的 WCF 抽象类实现。WsdlExporter 类型使用附加的策略表达式生成 WSDL 元数据。

若要导出自定义 WSDL 元数据或终结点行为的 WSDL 扩展、协定行为或服务终结点中的绑定元素,可以实现 IWsdlExportExtension 接口。在生成 WSDL 文档时,WsdlExporter 为实现 ServiceEndpoint 接口的绑定元素、操作行为、协定行为和终结点行为查看 IWsdlExportExtension 实例。

2. 发布服务元数据

WCF 服务通过公开一个或多个元数据终结点来发布元数据。使用标准协议(如 MEX 和 HTTP/GET 请求)发布服务元数据可使服务元数据变得可用。元数据终结点与其他服务终结点类似,都具有地址、绑定和协定。可以在配置中或在代码中将元数据终结点添加到服务主机。

若要发布 WCF 服务的元数据终结点,必须首先将 ServiceMetadataBehavior 服务行为的实例添加到服务。将 System. ServiceModel. Description. ServiceMetadataBehavior 实例添加到服务,使服务增添了通过公开一个或多个元数据终结点发布元数据的功能。添加 System. ServiceModel. Description. ServiceMetadataBehavior 服务行为后,可以公开支持 MEX 协议的元数据终结点或响应 HTTP/GET 请求的元数据终结点。

若要添加使用 MEX 协议的元数据终结点,请向使用名为 IMetadataExchange 的服务协定的服务主机添加服务终结点。WCF 定义 IMetadataExchange 具有此服务协定名称的接口。WS – MetadataExchange 终结点或 MEX 终结点可以使用由 MetadataExchangeBindings 类上静态工厂方法公开的四个默认绑定之一,以匹配 WCF 工具(如 Svcutil. exe)使用的默认绑定。

也可以使用自定义绑定配置 MEX 元数据终结点。

ServiceMetadataBehavior 使用 System.ServiceModel.Description.WsdlExporter 来导出服务中所有服务终结点的元数据。

通过将 ServiceMetadataBehavior 实例作为服务主机的扩展添加，ServiceMetadataExtension 增强了服务主机。System.ServiceModel.Description.ServiceMetadataExtension 提供了元数据发布协议的实现。还可以使用 System.ServiceModel.Description.ServiceMetadataExtension 通过访问 Metadata 属性来在运行时获取服务的元数据。

小心

如果在应用程序配置文件中添加 MEX 终结点，然后尝试在代码中向服务主机添加 ServiceMetadataBehavior，则会得到以下异常：

System.InvalidOperationException：在服务 Service1 实现的协定列表中找不到协定名称 "ImetadataExchange"。将 ServiceMetadataBehavior 添加到配置文件或直接添加到 ServiceHost，以启用对该协定的支持。

通过在配置文件中 ServiceMetadataBehavior 或在代码中同时添加终结点和 ServiceMetadataBehavior，可以解决此问题。

当发布公开两个不同服务协定（两个服务协定包含具有相同名称的操作）的服务的元数据时，会引发异常。例如，如果有一个服务公开了一个名为 ICarService 的服务协定，该服务协定具有一个 Get(Car c) 操作，且同一服务还公开了一个名为 IBookService 的服务协定，该服务协定具有一个 Get(Book b) 操作，则当生成该服务的元数据时，会引发异常或显示错误消息。若要解决此问题，请执行下列操作之一：

◇ 重命名其中的一项操作；

◇ 将 Name 设置为其他名称；

◇ 使用 Namespace 属性将其中一项操作的命名空间设置为其他命名空间。

3. 检索服务元数据

WCF 可以使用标准化协议（如 WS – MetadataExchange 和 HTTP）检索服务元数据。这两种协议均受 MetadataExchangeClient 类型支持。通过提供地址和可选绑定，使用 System.ServiceModel.Description.MetadataExchangeClient 类型检索服务元数据。由 System.ServiceModel.Description.MetadataExchangeClient 实例使用的绑定可以是 MetadataExchangeBindings 静态类中的默认绑定之一、用户提供的绑定或从 IMetadataExchange 协定的终结点配置加载的绑定。System.ServiceModel.Description.MetadataExchangeClient 也可以使用 HttpWebRequest 类型来解析 HTTP URL 对元数据的引用。

默认情况下，System.ServiceModel.Description.MetadataExchangeClient 实例与单个 ChannelFactoryBase 实例关联。通过重写 ChannelFactoryBase 虚拟方法，可以更改或替换由

System.ServiceModel.Description.MetadataExchangeClient 使用的 GetChannelFactory 实例。同样，通过重写 System.Net.HttpWebRequest 虚拟方法，可以更改或替换由 System.ServiceModel.Description.MetadataExchangeClient 使用的 MetadataExchangeClient.GetWebRequest 实例以发出 HTTP/GET 请求。

Svcutil.exe 工具可以检索服务元数据使用 Ws–metadataexchange 或 HTTP/GET 请求/target:metadata 开关和地址。Svcutil.exe 下载指定位置的元数据并将文件保存到磁盘。Svcutil.exe 在内部使用 System.ServiceModel.Description.MetadataExchangeClient 实例，并加载其名称与传递到 Svcutil.exe 的地址的方案（如果存在）匹配的 MEX 终结点配置（从应用程序配置文件）。否则，Svcutil.exe 默认使用由 MetadataExchangeBindings 静态工厂类型定义的绑定之一。

4. 导入服务元数据

在 WCF 中，元数据导入是从服务的元数据生成服务或其组成部分的抽象表示的过程。例如，WCF 可以从服务的 WSDL 文档导入 ServiceEndpoint 实例、Binding 实例或 ContractDescription 实例。若要在 WCF 中导入服务元数据，请使用 MetadataImporter 抽象类的实现。派生自 System.ServiceModel.Description.MetadataImporter 类的类型实现对导入元数据格式的支持，这些元数据格式利用了 WCF 中的 WS–Policy 导入逻辑。

System.ServiceModel.Description.MetadataImporter 实现收集附加到 PolicyConversionContext 对象中服务元数据的策略表达式。然后，通过在 System.ServiceModel.Description.MetadataImporter 属性中调用 IPolicyImportExtension 接口的实现，PolicyImportExtensions 在导入元数据的过程中处理策略。

通过将自己的 System.ServiceModel.Description.MetadataImporter 接口实现添加到 IPolicyImportExtension 实例上的 PolicyImportExtensions 集合，可以添加对将新策略断言导入到 System.ServiceModel.Description.MetadataImporter 的支持。或者，可以在客户端应用程序配置文件中注册策略导入扩展。

System.ServiceModel.Description.WsdlImporter 类型是 System.ServiceModel.Description.MetadataImporter 所包含的 WCF 抽象类的实现。System.ServiceModel.Description.WsdlImporter 类型可以导入含有附加策略（这些策略捆绑在 MetadataSet 对象中）的 WSDL 元数据。

通过实现 IWsdlImportExtension 接口，然后将实现添加到 WsdlImportExtensions 实例上的 System.ServiceModel.Description.WsdlImporter 属性，可以添加对导入 WSDL 扩展的支持。System.ServiceModel.Description.WsdlImporter 还可以加载在客户端应用程序配置文件中注册的 System.ServiceModel.Description.IWsdlImportExtension 接口的实现。

5. 动态绑定

如果终结点的绑定更改,或者希望为使用相同协定但具有不同绑定的终结点创建一个通道,则可以动态更新用来为服务终结点创建通道的绑定。可以使用 MetadataResolver 静态类在运行时为实现特定协定的服务终结点检索和导入元数据。然后可以使用导入的 System.ServiceModel.Description.ServiceEndpoint 对象为所需终结点创建客户端或通道工厂。

5.2 导出和导入元数据

在 Windows Communication Foundation(WCF)中,导出元数据是一个描述服务终结点并将其映射到一个并行的标准化表示形式的过程,客户端可以使用这种表示形式来了解如何使用服务。导入服务元数据是一个从服务元数据生成 ServiceEndpoint 实例或部分的过程。

1. 导出元数据

若要从 System.ServiceModel.Description.ServiceEndpoint 实例导出元数据,请使用 MetadataExporter 抽象类的实现。WsdlExporter 类型是 MetadataExporter 所包含的 WCF 抽象类的实现。

System.ServiceModel.Description.WsdlExporter 类型使用封装在 MetadataSet 实例中的附加策略表达式生成 Web 服务描述语言(WSDL)元数据。可以使用 System.ServiceModel.Description.WsdlExporter 实例以迭代方式为 ContractDescription 对象和 ServiceEndpoint 对象导出元数据。还可以导出 ServiceEndpoint 对象的集合,并将其与特定的服务名称相关联。

备注

只可使用 WsdlExporter 从包含公共运行库(CLR)类型信息的 ContractDescription 实例中导出元数据,例如使用 ContractDescription 方法创建的 ContractDescription.GetContract 实例,或作为 ServiceDescription 实例 ServiceHost 的一部分创建的实例。对于从服务元数据导入的或构建时没有类型信息的 WsdlExporter 实例,不能使用 ContractDescription 从此类实例导出元数据。

2. 导入元数据

(1)导入 WSDL 文档

若要在 WCF 中导入服务元数据,请使用 MetadataImporter 抽象类的实现。System.ServiceModel.Description.WsdlImporter 类型是 MetadataImporter 所包含的 WCF 抽象类的实

现。WsdlImporter 类型使用捆绑在 MetadataSet 对象中的附加策略来导入 WSDL 元数据。

WsdlImporter 类型可以让用户控制如何导入元数据。可以导入所有终结点、所有绑定或所有协定。可以导入与特定的 WSDL 服务、绑定或端口类型相关联的所有终结点。还可以导入特定的 WSDL 端口的终结点、特定的 WSDL 绑定的绑定或特定的 WSDL 端口类型的协定。

WsdlImporter 还公开 KnownContracts 属性，可让用户指定一组不需要导入的协定。WsdlImporter 使用 KnownContracts 属性中的协定，而不是从元数据导入具有相同限定名的协定。

（2）导入策略

WsdlImporter 类型收集附加到消息、操作和终结点策略主题的策略表达式，然后使用 IPolicyImportExtension 集合中的 PolicyImportExtensions 实现导入策略表达式。

策略导入逻辑自动处理对同一 WSDL 文档中的策略表达式的策略引用，并使用 wsu:Id 或 xml:id 属性进行标识。策略导入逻辑通过将策略表达式的大小限制为 4 096 个节点，从而防止应用程序进行循环策略应用，此处的节点是以下元素之一：wsp:Policy、wsp:All、wsp:ExactlyOne 和 wsp:policyReference。

策略导入逻辑还自动标准化策略表达式。嵌套的策略表达式和 wsp:Optional 属性不进行标准化。完成的标准化处理的数量限制为 4 096 步，其中每一步都会产生一个策略断言（即 wsp:ExactlyOne 元素的子元素）。

WsdlImporter 类型最多尝试 32 种附加到不同 WSDL 策略主题的备用策略组合。如果没有完全导入任何组合，则使用第一个组合来构造部分自定义绑定。

（3）错误处理

MetadataExporter 和 MetadataImporter 类型都公开 Errors 属性，该属性可以包含一个在导出和导入过程（可在实现工具时使用）中分别出现的错误和警告消息的集合。

WsdlImporter 类型通常会为在导入过程中捕获的异常引发一个异常，并且将相应的错误添加到其 Errors 属性中。但是 ImportAllContracts、ImportAllBindings、ImportAllEndpoints 和 ImportEndpoints 方法不会引发这些异常，因此，必须检查 Errors 属性以确定在调用这些方法时是否发生了任何问题。

WsdlExporter 类型会再次引发在导出过程中捕获的所有异常。这些异常不会作为 Errors 属性中的错误而捕获。一旦 WsdlExporter 引发异常，则会进入错误状态，并且无法重用。当某个操作因使用了通配符操作而无法导出时以及在遇到重复的绑定名称时，WsdlExporter 确实会将警告添加到其 Errors 属性中。

5.2.1 元数据导入服务终结点

将元数据导入服务终结点

（1）声明 EndpointAddress 对象，并使用服务的元数据交换（MEX）地址的统一资源标识符（URI）对其进行初始化。

```
EndpointAddress mexAddress = new EndpointAddress ( " http://localhost:8000/
ServiceModelSamples/service/mex");
```

（2）创建 MetadataExchangeClient、传入 MEX 地址并调用 GetMetadata。此操作将从服务中检索元数据。

```
MetadataExchangeClient mexClient = new MetadataExchangeClient(mexAddress);
mexClient.ResolveMetadataReferences = true;
MetadataSet metaSet = mexClient.GetMetadata();
```

（3）创建 WsdlImporter、传入先前检索的元数据并调用 ImportAllContracts。此操作将生成 ContractDescription 对象的集合。还可以根据需要调用 ImportAllEndpoints 或 ImportAllBindings。

```
WsdlImporter importer = new WsdlImporter(metaSet);
System.Collections.ObjectModel.Collection<ContractDescription> contracts
= importer.ImportAllContracts();
```

备注

导入元数据后，将无法创建客户端通道或导出元数据。这是因为此时没有可用的类型信息。在实际与服务进行交互或导出元数据时需要类型信息。若要生成类型信息，需要生成代码，如以下步骤（4）和（5）所示。或者可以使用 MetadataResolver 帮助器类。

（4）为每个协定生成类型信息。

```
ServiceContractGenerator generator = new ServiceContractGenerator();
foreach (ContractDescription contract in contracts)
{
    generator.GenerateServiceContractType(contract);
}

        if (generator.Errors.Count != 0)
            throw new Exception("There were errors during code compilation.");
```

（5）现在，可以使用此信息。下面的示例生成 C# 源代码。

```
System.CodeDom.Compiler.CodeGeneratorOptions options = new System.CodeDom.
Compiler.CodeGeneratorOptions();
options.BracingStyle = "C";
System.CodeDom.Compiler.CodeDomProvider codeDomProvider = System.CodeDom.
Compiler.CodeDomProvider.CreateProvider("C#");
System.CodeDom.Compiler.IndentedTextWriter textWriter = new System.CodeDom.
Compiler.IndentedTextWriter(new System.IO.StreamWriter(outputFile));
```

```
codeDomProvider.GenerateCodeFromCompileUnit(generator.TargetCompileUnit,
textWriter, options);
textWriter.Close();
```

5.2.2 服务终结点导出元数据

从服务终结点导出元数据

（1）创建一个新的 Visual Studio 控制台应用程序项目。在生成的 Program.cs 文件的 main() 方法中添加下列步骤所示的代码。

（2）创建 WsdlExporter。

```
WsdlExporter exporter = new WsdlExporter();
```

（3）将 PolicyVersion 属性设置为 PolicyVersion 枚举值之一。此示例将该值设置为与 WS-Policy 1.5 对应的 Policy15。

```
exporter.PolicyVersion = PolicyVersion.Policy15;
```

（4）创建 ServiceEndpoint 对象的数组。

```
ServiceEndpoint[] myServiceEndpoints = new ServiceEndpoint[2];
ContractDescription myDescription = new ContractDescription("myContract");
myServiceEndpoints[0] = new ServiceEndpoint(myDescription, new BasicHttpBinding(),new EndpointAddress("http://localhost/myservice"));
myServiceEndpoints[1] = new ServiceEndpoint(myDescription, new BasicHttpBinding(),new EndpointAddress("http://localhost/myservice"));
```

（5）为每个服务终结点导出元数据。

```
foreach(ServiceEndpoint endpoint in myServiceEndpoints)
{ exporter.ExportEndpoint(endpoint);}
```

（6）检查以确保在导出过程中不会发生任何错误，并检索元数据。

```
MetadataSet metadataDocs = null;
if(exporter.Errors.Count != 0)
{ metadataDocs = exporter.GetGeneratedMetadata();}
```

（7）现在可以使用元数据，例如通过调用 WriteTo(XmlWriter) 方法将它写入文件。

下面列出了此示例的完整代码。

```
using System; using System.ServiceModel;
using System.ServiceModel.Description;
namespace WsdlExporterSample
{ class Program
  { static void Main(string[] args)
    { WsdlExporter exporter = new WsdlExporter();
```

```
            exporter.PolicyVersion = PolicyVersion.Policy15;
            ServiceEndpoint[] myServiceEndpoints = new ServiceEndpoint[2];
        ContractDescription myDescription = new ContractDescription("myContract");
            myServiceEndpoints[0] = new ServiceEndpoint(myDescription, new BasicHttpBinding(),new EndpointAddress("http://localhost/myservice"));
    myServiceEndpoints[1] = new ServiceEndpoint(myDescription, new BasicHttpBinding(),new EndpointAddress("http://localhost/myservice"));
            foreach(ServiceEndpoint endpoint in myServiceEndpoints)
            { exporter.ExportEndpoint(endpoint);}
            MetadataSet metadataDocs = null;
            if(exporter.Errors.Count! = 0)
{ metadataDocs = exporter.GetGeneratedMetadata();}}}}
```

5.2.3 ServiceDescription 和 WSDL 引用

1. ServiceDescription 如何映射到 WSDL 1.1

可以使用 WCF 为服务从 ServiceDescription 实例导出 WSDL 文档。发布元数据终结点时,将为服务自动生成 WSDL 文档。

还可以使用 ServiceEndpoint 类型从 WSDL 文档导入 ContractDescription 实例、Binding 实例和 WsdlImporter 实例。

由 WCF 导出的 WSDL 文档将从外部 XML 架构文档导入使用的所有 XML 架构定义。为服务中数据类型使用的每个目标命名空间导出单独的 XML 架构文档。同样,为服务协定使用的每个目标命名空间导出单独的 WSDL 文档。

2. ServiceDescription

ServiceDescription 实例映射到 wsdl:service 元素。ServiceDescription 实例包含 ServiceEndpoint 实例的集合,其中每个实例都映射到单独的 wsdl:port 元素。见表 5-1。

表 5-1　ServiceDescription 实例的 WSDL 映射

属性	WSDL 映射
Name	wsdl:service /@ name 服务的值
Namespace	服务的 wsdl:service 定义的 targetNamespace
Endpoints	服务的 wsdl:port 定义

3. ServiceEndpoint

ServiceEndpoint 实例映射到 wsdl:port 元素。ServiceEndpoint 实例包含一个地址、绑定和协定。见表 5-2。

实现 IWsdlExportExtension 接口的终结点行为可以修改它们所附加到的终结点的 wsdl:port 元素。

表 5-2　ServiceDescription 实例的 WSDL 映射

属性	WSDL 映射
Name	wsdl:port /@ name 终结点的值与'wsdl:binding'/@ name 终结点绑定的值
Address	终结点的 wsdl:port 定义的地址。 终结点的传输确定地址的格式。例如,对于 WCF 支持的传输,则可能是 SOAP 地址或终结点引用
Binding	终结点的 wsdl:binding 定义。 与 wsdl:binding 定义不同,WCF 中的绑定不会与任何一个协定关联
Contract	终结点的 wsdl:portType 定义
Behaviors	实现 IWsdlExportExtension 接口的终结点行为可以修改终结点的 wsdl:port

4. 绑定

ServiceEndpoint 实例的绑定实例映射到 wsdl:binding 定义。见表 5-3。wsdl:binding 定义与 wsdl:portType 绑定不同,前者必须与特定的 WCF 定义相关联,而后者独立于任何协定。

绑定由绑定元素的集合组成。每个元素描述终结点与客户端的通信方式的某一方面。另外,绑定还具有一个 MessageVersion,指示终结点的 EnvelopeVersion 和 AddressingVersion。

表 5-3　ServiceEndpoint 实例的 WSDL 映射

属性	WSDL 映射
Name	在终结点的默认名称中使用,该名称是以下画线分隔追加的协定名称的绑定名称
Namespace	targetNamespace 定义的 wsdl:binding。 导入时,如果将策略附加到 WSDL 端口,则导入的绑定命名空间将映射到 targetNamespace 定义的 wsdl:port

表 5-3(续)

属性	WSDL 映射
BindingElementCollection,由 CreateBindingElements( )方法返回	wsdl:binding 定义的各种域特定的扩展,通常是策略断言
MessageVersion	终结点的 EnvelopeVersion 和 AddressingVersion。如果指定 MessageVersion.None,则 WSDL 绑定不包含 SOAP 绑定,并且 WSDL 端口不包含 WS-Addressing 内容。该设置通常用于 Plain Old XML (POX)终结点

BindingElements:终结点绑定的绑定元素映射到 wsdl:binding 中的各种 WSDL 扩展,如策略断言。

绑定的 TransportBindingElement 为 SOAP 绑定确定传输统一资源标识符(URI)。

AddressingVersion:绑定上的 AddressingVersion 映射到 wsd:port 中使用的寻址版本。WCF 支持 SOAP 1.1 和 SOAP 1.2 地址以及 WS-Addressing 08/2004 和 WS-Addressing 1.0 终结点引用。

EnvelopeVersion:绑定上的 EnvelopeVersion 映射到 wsdl:binding 中使用的 SOAP 的版本。WCF 支持 SOAP 1.1 和 SOAP 1.2 绑定。

5. 协定

ContractDescription 实例的 ServiceEndpoint 实例映射到 wsdl:portType。ContractDescription 实例描述给定协定的所有操作。见表 5-4。

表 5-4 ContractDescription 实例的 WSDL 映射

属性	WSDL 映射
Name	wsdl:portType /@ name 协定的值
Namespace	wsdl:portType 定义的 targetNamespace
SessionMode	wsdl:portType /@ msc:usingSession 协定的值。此属性是 WSDL 1.1 的 WCF 扩展
Operations	协定的 wsdl:operation 定义

6. 操作

OperationDescription 实例映射到 wsdl:portType / wsdl:operation。见表 5-5。OperationDescription 包含用于描述操作消息的 MessageDescription 实例的集合。

如下两个操作行为广泛地参与 OperationDescription 到 WSDL 文档的映射方式:

DataContractSerializerOperationBehavior 和 XmlSerializerOperationBehavior。

表 5-5　OperationDescription 实例的 WSDL 映射

属性	WSDL 映射
Name	wsdl:portType / wsdl:operation /@ name 操作的值
ProtectionLevel	附加到此操作的 wsdl:binding/wsdl:operation 消息的安全策略中的保护断言
IsInitiating	wsdl:portType / wsdl:operation /@ msc:isInitiating 操作的值。此属性是 WSDL 1.1 的 WCF 扩展
IsTerminating	wsdl:portType / wsdl:operation /@ msc:isTerminating 操作的值。此属性是 WSDL 1.1 的 WCF 扩展
Messages	wsdl:portType / wsdl:operation / wsdl:input 和 wsdl:portType/ wsdl:operation / wsdl:output 操作的消息
Faults	wsdl:portType / wsdl:operation / wsdl:fault 定义,则为该操作
Behaviors	DataContractSerializerOperationBehavior 和 XmlSerializerOperationBehavior 处理操作绑定和操作消息

7. DataContractSerializerOperationBehavior

操作的 DataContractSerializerOperationBehavior 是用于为该操作导出 WSDL 消息和绑定的 IWsdlExportExtension 实现。见表 5-6。XML 架构类型是使用 XsdDataContractExporter 导出的。DataContractSerializerOperationBehavior 还为该操作确定用途、样式和要使用的架构导出程序与导入程序。

表 5-6　DataContractSerializerOperationBehavior 的 WSDL 映射

属性	WSDL 映射
DataContractFormatAttribute	Style 此特性的属性将映射到 wsdl:binding / wsdl:operation / soap:operation /@ style 操作的值。 DataContractSerializerOperationBehavior 仅支持 WSDL 中架构类型的文字用法

8. XmlSerializerOperationBehavior

操作的 XmlSerializerOperationBehavior 是用于为该操作导出 WSDL 消息和绑定的

IWsdlExportExtension 实现。见表 5-7。XML 架构类型是使用 XmlSchemaExporter 导出的。XmlSerializerOperationBehavior 还为该操作确定用途、样式和要使用的架构导出程序与导入程序。

表 5-7 XmlSerializerOperationBehavior 的 WSDL 映射

属性	WSDL 映射
XmlSerializerFormatAttribute	Style 此特性的属性将映射到 wsdl: binding / wsdl: operation / soap: operation /@ style 操作的值。 Use 此特性的属性将映射到 wsdl: binding / wsdl: operation / soap: operation/ /@ use 操作中的所有消息的值

9. 消息

AMessageDescription 实例映射到 wsdl: message 引用 wsdl: portType / wsdl: operation / wsdl: input 或 wsdl: portType / wsdl: operation / wsdl: output 在操作中的消息。MessageDescription 包含正文和标头。见表 5-8。

表 5-8 AMessageDescription 实例的 WSDL 映射

属性	WSDL 映射
Action	消息的 SOAP 或 WS-Addressing 操作。 请注意,使用操作字符串""的操作不用 WSDL 表示
Direction	MessageDirection. Input 映射到 wsdl: input。 MessageDirection. Output 映射到 wsdl: output
ProtectionLevel	附加到此消息的 wsdl: message 定义的安全策略中的保护断言
Body	消息的消息正文
Headers	消息的标头
ContractDescription. Name, OperationContract. Name	在导出时,用于派生 wsdl: message/@ name 值

(1) 消息正文

AMessageBodyDescription 实例映射到 wsdl: message / wsdl: part 的消息的正文定义。消息正文可以包装也可以裸露。见表 5-9。

表 5-9 AMessageBodyDescription 实例的 WSDL 映射

属性	WSDL 映射
WrapperName	如果样式不是 RPC，则 WrapperName 映射到引用的元素名称 wsdl:message/wsdl:part 与@ name 设置为"parameters"
WrapperNamespace	如果样式不是 RPC，则 WrapperNamespace 映射到元素命名空间 wsdl:message/wsdl:part 与@ name 设置为"parameters"
Parts	此消息正文的消息部分
ReturnValue	如果存在包装元素（文档包装样式或 RPC 样式），否则包装元素的子元素的第一个 wsdl:message / wsdl:part 消息中

（2）消息部分

AMessagePartDescription 实例映射到 wsdl:message / wsdl:part 和 XML 架构类型或消息部分所指向的元素。见表 5-10。

表 5-10 AMessagePartDescription 实例的 WSDL 映射

属性	WSDL 映射
Name	wsd:message / wsdl:part /@ name 以及消息部分所指向的元素的名称的值
Namespace	消息部分所指向的元素的命名空间
Index	索引 wsdl:message / wsdl:part 消息
ProtectionLevel	附加到此消息部分的 wsdl:message 定义的安全策略中的保护断言。将策略参数化，使其指向特定的消息部分
MessageType	消息部分所指向的元素的 XML 架构类型

（3）消息头

MessageHeaderDescription 实例是同时还映射到消息部分的 soap:header 绑定的消息部分。

（4）错误

AFaultDescription 实例映射到 wsdl:portType / wsdl:operation / wsdl:fault 定义和其关联 wsdl:message 定义。见表 5-11。将 wsdl:message 添加到与其关联的 WSDL 端口类型相同的目标命名空间。wsdl:message 有一个名为"详细信息"的消息部分，该部分指向对应于 DefaultType 实例的 FaultDescription 属性值的 XML 架构元素。

表 5 – 11　AFaultDescription 实例的 WSDL 映射

属性	WSDL 映射
Name	wsdl:portType / wsdl:operation / wsdl:fault /@ name 错误的值
Namespace	错误详细消息部分所指向的 XML 架构元素的命名空间
Action	错误的 SOAP 或 WS – Addressing 操作
ProtectionLevel	附加到此错误的 wsdl:message 定义的安全策略中的保护断言
DetailType	详细消息部分所指向的元素的 XML 架构类型
Name，ContractDescription.Name，OperationDescription.Name	用于派生 wsdl:message/@ name 错误消息的值

5.2.4　Svcutil.exe 将元数据从已编译的服务代码中导出

Svcutil.exe 可以导出已编译程序集中的服务、协定和数据类型的元数据，如下所示：

➢ 若要使用 Svcutil.exe 为一组程序集导出所有已编译服务协定的元数据，请将这些程序集指定为输入参数。这是默认行为。

➢ 若要使用 Svcutil.exe 导出已编译服务的元数据，请将该（这些）服务程序集指定为输入参数。必须使用 /serviceName 选项来指示要导出的服务的配置名称。Svcutil.exe 自动为指定的可执行程序集加载配置文件。

➢ 若要导出一组程序集内的所有数据协定类型，请使用 /dataContractOnly 选项。

备注

请使用 /reference 选项来指定所有相关程序集的文件路径。

导出已编译服务协定的元数据：

（1）将服务协定实现编译为一个或多个类库。

（2）在已编译程序集上运行 Svcutil.exe。

备注

您可能需要使用 /reference 开关指定任意相关程序集的文件路径。

`svcutil.exe Contracts.dll`

1. 导出已编译服务的元数据

（1）将服务实现编译为可执行程序集。

（2）为服务可执行程序集创建一个配置文件，并添加服务配置。

`<? xml version = "1.0" encoding = "utf - 8" ? >`

`< configuration >`

```
<system.serviceModel>
  <services>
    <service name = "MyService">
      <endpoint address = "finder" contract = "IPeopleFinder" binding = "wsHttpBinding" />
    </service>
  </services>
</system.serviceModel>
</configuration>
```

(3) 在使用 /serviceName 开关指定服务的配置名称的已编译服务可执行程序集上运行 Svcutil.exe。

备注

您可能需要使用 /reference 开关指定任意相关程序集的文件路径。

svcutil.exe /serviceName:MyService Service.exe /reference:path/Contracts.dll

2. 导出已编译数据协定的元数据

(1) 将数据协定实现编译为一个或多个类库。

(2) 在使用 /dataContract 开关指定应仅生成数据协定元数据的已编译程序集上运行 Svcutil.exe。

备注

您可能需要使用 /reference 开关指定任意相关程序集的文件路径。

svcutil.exe /dataContractOnly Contracts.dll

3. 导出服务协定的元数据。

svcutil.exe Contracts.dll

导出数据协定的元数据。

svcutil.exe /dataContractOnly Contracts.dll

导出服务实现的元数据。

svcutil.exe /serviceName:MyService Service.exe /reference:<path>/Contracts.dll

<path> 是 Contracts.dll 的路径。

```
[ServiceContract(ConfigurationName = "IPeopleFinder")]
public interface IPersonFinder
{ [OperationContract]  Address GetAddress(Person s); }
[DataContract]
public class Person
```

```
{   [DataMember]      public string firstName;
    [DataMember]      public string lastName;
    [DataMember]      public int age;  }
[DataContract]
public class Address
{   [DataMember]      public string city;
    [DataMember]      public string state;
    [DataMember]      public string street;
    [DataMember]      public int zipCode;
    [DataMember]      public Person person;  }
[ServiceBehavior(ConfigurationName = "MyService")]
public class MyService : IPersonFinder
{   public Address GetAddress(Person person)
    {   Address address = new Address();
        address.person = person;
        return address;       }    }
<?xml version = "1.0" encoding = "utf-8"?>
<configuration>
  <system.serviceModel>
    <services>
      <service name = "MyService">
        <endpoint address = "finder"
                binding = "basicHttpBinding"
                contract = "IPeopleFinder"/>
      </service>
    </services>
  </system.serviceModel>
</configuration>
```

5.3　发布元数据

　　Windows Communication Foundation（WCF）服务通过发布一个或多个元数据终结点来发布元数据。发布服务元数据之后,可以通过标准协议(如 WS – MetadataExchange（MEX）和 HTTP/GET 请求)来使用该元数据。元数据终结点类似于其他服务终结点,因为它们都有一个地址、一个绑定和一个协定,并且它们都可通过配置或命令代码添加到服务主机。

1. 发布元数据终结点

若要发布 WCF 服务的终结点，首先必须将 ServiceMetadataBehavior 服务行为添加到该服务。添加一个 System.ServiceModel.Description.ServiceMetadataBehavior 实例将允许服务公开元数据终结点。添加 System.ServiceModel.Description.ServiceMetadataBehavior 服务行为之后，就可以公开支持 MEX 协议或响应 HTTP/GET 请求的元数据终结点。

System.ServiceModel.Description.ServiceMetadataBehavior 使用 WsdlExporter 来导出服务中所有服务终结点的元数据。

System.ServiceModel.Description.ServiceMetadataBehavior 添加一个 ServiceMetadataExtension 实例作为服务主机的扩展。System.ServiceModel.Description.ServiceMetadataExtension 提供了元数据发布协议的实现。还可以使用 System.ServiceModel.Description.ServiceMetadataExtension 通过访问 ServiceMetadataExtension.Metadata 属性来在运行时获取服务的元数据。

2. MEX 元数据终结点

要添加使用 MEX 协议的元数据终结点，请将服务终结点添加到使用 IMetadataExchange 服务协定的服务主机。WCF 包括 IMetadataExchange 接口以及此服务协定名称，可以将此名称用作 WCF 编程模型的一部分。WS – MetadataExchange 终结点，即 MEX 终结点，可以使用静态工厂方法在 MetadataExchangeBindings 类中公开的四种默认绑定之一来匹配 Svcutil.exe 之类的 WCF 工具所使用的默认绑定。还可以使用自定义绑定来配置 MEX 元数据终结点。

3. HTTP GET 元数据终结点

若要将元数据终结点添加到响应 HTTP/GET 请求的服务，请将 HttpGetEnabled 的 System.ServiceModel.Description.ServiceMetadataBehavior 属性设置为 true。将 HttpsGetEnabled 的 System.ServiceModel.Description.ServiceMetadataBehavior 属性设置为 true，还可以配置使用 HTTPS 的元数据终结点。

5.3.1 使用配置文件发布服务的元数据

有两种方式可以指定服务应如何发布元数据：使用配置文件和使用代码。本节讲述如何使用配置文件发布服务的元数据。

小心

本节讲述如何以不安全的方式发布元数据。任何客户端都可以检索服务的元数据。如果要求服务以安全方式发布元数据，请参阅自定义安全元数据终结点。

通过发布元数据，客户端可以使用 WS – Transfer GET 请求或使用？wsdl 查询字符串的 HTTP/GET 来检索元数据。若要确保代码能够工作，必须创建一个基本的 WCF 服务。为为了简单起见,将在以下代码中提供一个基本的自承载服务。

```csharp
using System; using System.Runtime.Serialization; using System.ServiceModel;
using System.ServiceModel.Description;
namespace Metadata.Samples
{
    [ServiceContract]
    public interface ISimpleService
    {
        [OperationContract]
        string SimpleMethod(string msg);
    }
    class SimpleService : ISimpleService
    {   public string SimpleMethod(string msg)
        {
            Console.WriteLine("The caller passed in " + msg);
            return "Hello " + msg;
        }
    }
    class Program
    {
        static void Main(string[] args)
        {
            ServiceHost host = new ServiceHost(typeof(SimpleService),
                new Uri("http://localhost:8001/MetadataSample"));
            try
            {   host.Open();
                Console.WriteLine("The service is ready.");
                Console.WriteLine("Press <ENTER> to terminate service.");
                Console.WriteLine();
                Console.ReadLine();
                host.Close();  }
            catch (CommunicationException commProblem)
            {   Console.WriteLine(" There was a communication problem. " +
commProblem.Message);  Console.Read();    }   }   }
}
```

此服务为自承载服务,并且是使用配置文件配置的。下面的配置文件作为开始点。

```xml
<configuration>
  <system.serviceModel>
    <services>
      <service name="Metadata.Example.SimpleService">
```

```
       <endpoint address = " "
              binding = "basicHttpBinding"
              contract = "Metadata.Example.ISimpleService" />
    </service>
  </services>
  <behaviors>
  </behaviors>
</system.serviceModel>
</configuration>
```

1. 使用应用程序配置文件发布 WCF 服务的元数据

(1)在 App.config 文件内,在结束 </services> 元素之后创建 <behaviors> 元素。

(2)在 <behaviors> 元素内,添加一个 <serviceBehaviors> 元素。

(3)向 <behavior> <serviceBehaviors> 元素中添加一个 name 元素,并为元素的 <behavior> 属性指定一个值。

(4)向 <serviceMetadata> 元素中添加一个 <behavior> 元素。将 httpGetEnabled 属性设置为 true,并将 policyVersion 属性设置为 Policy15。httpGetEnabled 可让服务响应 HTTP GET 请求发出的元数据请求。policyVersion 通知服务在生成元数据时应符合 WS – Policy 1.5。

(5)将 behaviorConfiguration 特性添加到 <service> 元素,并指定在步骤(1)中添加的 name 元素的 <behavior> 特性,如下面的代码示例所示。

```
<services>
  <service
    name = "Metadata.Example.SimpleService"
    behaviorConfiguration = "SimpleServiceBehavior">
    ...
  </service>
</services>
<behaviors>
  <serviceBehaviors>
    <behavior name = "SimpleServiceBehavior">
      <serviceMetadata httpGetEnabled = "True" policyVersion = "Policy15" />
    </behavior>
  </serviceBehaviors>
</behaviors>
```

(6)添加一个或多个其协定设置为 <endpoint> 的 IMetadataExchange 元素,如下面的代码示例所示。

```
<services>
  <service
    name = "Metadata.Example.SimpleService"
    behaviorConfiguration = "SimpleServiceBehavior" >
   <endpoint address = ""
        binding = "wsHttpBinding"
        contract = "Metadata.Example.ISimpleService" />
   <endpoint address = "mex"
        binding = "mexHttpBinding"
        contract = "IMetadataExchange" />
  </service>
</services>
```

（7）对于上一步中添加的元数据终结点，将 binding 特性设置为下列值之一：
◇ mexHttpBinding（对于 HTTP 发布）；
◇ mexHttpsBinding（对于 HTTPS 发布）；
◇ mexNamedPipeBinding（对于命名管道发布）；
◇ mexTcpBinding（对于 TCP 发布）。

（8）对于上一步中添加的元数据终结点，将地址设置为等于：
◇ 一个空字符串，以在基址和元数据绑定相同的情况下将主机应用程序的基址用作发布点。
◇ 一个相对地址，前提是主机应用程序具有一个基址。
◇ 一个绝对地址。

（9）生成并运行控制台应用程序。

（10）使用 Internet Explorer 浏览服务的基址（本示例中的 http://localhost:8001/MetadataSample）并验证是否已打开元数据发布。如果未显示上述内容，则结果页顶部会显示："Metadata publishing for this service is currently disabled"（当前禁用服务的元数据发布）。

2. 使用默认终结点

若要配置使用默认终结点的服务上的元数据，请在上一个示例中的配置文件中指定 ServiceMetadataBehavior，但不要指定任何终结点。配置文件将类似如下所示。

```
<configuration>
  <system.serviceModel>
   <behaviors>
     <serviceBehaviors>
       <behavior name = "SimpleServiceBehavior">
        <serviceMetadata httpGetEnabled = "True" policyVersion = "Policy12" >
```

            </behavior>
          </serviceBehaviors>
        </behaviors>
      </system.serviceModel>
    </configuration>

由于该服务有一个 ServiceMetadataBehavior 设置为 httpGetEnabled 的 true,因此该服务启用了发布元数据,但是由于未显式添加任何终结点,因此运行时添加默认终结点。有关以下内容的详细信息默认终结点、绑定和行为,请参阅简化配置和简化配置 WCF 服务。

3. 示例

下面的代码示例演示基本 WCF 服务的实现,以及发布服务的元数据的配置文件。

```csharp
using System;using System.Runtime.Serialization;using System.ServiceModel;
using System.ServiceModel.Description;
namespace Metadata.Samples
{
    [ServiceContract]
    public interface ISimpleService
    {
        [OperationContract]
        string SimpleMethod(string msg);
    }
    class SimpleService : ISimpleService
    {
        public string SimpleMethod(string msg)
        {
            Console.WriteLine("The caller passed in " + msg);
            return "Hello " + msg;
        }
    }
    class Program
    {
        static void Main(string[] args)
        {
            ServiceHost host = new ServiceHost(typeof(SimpleService),
                new Uri("http://localhost:8001/MetadataSample"));
            try
            { host.Open();
```

```
                Console.WriteLine("The service is ready.");
                Console.WriteLine("Press <ENTER> to terminate service.");
                Console.WriteLine();
                Console.ReadLine();
                host.Close();   }
            catch (CommunicationException commProblem)
            {
                Console.WriteLine("There was a communication problem. " +
commProblem.Message);
                Console.Read();
            }    }    }  }
```

配置文件：

```
<configuration>
  <system.serviceModel>
    <behaviors>
      <serviceBehaviors>
        <behavior name = "SimpleServiceBehavior">
          <serviceMetadata httpGetEnabled = "True" policyVersion = "Policy12" />
          <serviceDebug includeExceptionDetailInFaults = "False" />
        </behavior>
      </serviceBehaviors>
    </behaviors>
  </system.serviceModel>
</configuration>
```

5.3.2 使用代码发布服务的元数据

小心

本节讲述如何以不安全的方式发布元数据。任何客户端都可以检索服务的元数据。

通过发布元数据，客户端可以使用 WS – Transfer GET 请求或使用？wsdl 查询字符串的 HTTP/GET 来检索元数据。若要确保代码能够工作，必须创建一个基本的 WCF 服务。在以下代码中提供了一个基本的自承载服务。

```
using System;
using System.Runtime.Serialization;
using System.ServiceModel;
using System.ServiceModel.Description;
namespace Metadata.Samples
```

第5章 元 数 据

```
    [ServiceContract]
public interface ISimpleService
{
    [OperationContract]
    string SimpleMethod(string msg);    }
class SimpleService : ISimpleService
{
    public string SimpleMethod(string msg)
    {
        Console.WriteLine("The caller passed in " + msg);
        return "Hello " + msg;
    }   }
```

在代码中发布元数据的步骤如下：

(1)在控制台应用程序的主方法中,通过传入服务类型和基址来实例化 ServiceHost 对象。

```
ServiceHost svcHost = new ServiceHost(typeof(SimpleService), new Uri("http://localhost:8001/MetadataSample"));
```

(2)紧接在步骤(1)的代码下面创建一个 try 块,这将捕获在服务运行过程中引发的任何异常。

```
try{}
catch(CommunicationException commProblem)
{
    Console.WriteLine("There was a communication problem. " + commProblem.Message);
    Console.Read();
}
```

(3)检查服务主机是否已经包含一个 ServiceMetadataBehavior,如果没有,则创建一个新的 ServiceMetadataBehavior 实例 ServiceMetadataBehavior smb = svcHost.Description.Behaviors.Find<ServiceMetadataBehavior>();

```
if(smb == null)
    smb = new ServiceMetadataBehavior();
```

(4)将 HttpGetEnabled 属性设置为 true。

```
smb.HttpGetEnabled = true;
```

(5)ServiceMetadataBehavior 包含一个 MetadataExporter 属性。MetadataExporter 包含一个 PolicyVersion 属性。将 PolicyVersion 属性的值设置为 Policy15。还可以将 PolicyVersion 属性设置为 Policy12。当设置为 Policy15 时,元数据导出程序会使用符合 WS – Policy 1.5 的

元数据生成策略信息。当设置为 Policy12 时,元数据导出程序会生成符合 WS – Policy 1.2 的策略信息。

```
smb.MetadataExporter.PolicyVersion = PolicyVersion.Policy15;
```

(6) 将 ServiceMetadataBehavior 实例添加到服务主机的行为集合。

```
svcHost.Description.Behaviors.Add(smb);
```

(7) 将元数据交换终结点添加到服务主机。

```
svcHost.AddServiceEndpoint(
  ServiceMetadataBehavior.MexContractName,
  MetadataExchangeBindings.CreateMexHttpBinding(),
  "mex");
```

(8) 将应用程序终结点添加到服务主机。

```
svcHost.AddServiceEndpoint(typeof(ISimpleService), new WSHttpBinding(), "");
```

备注

如果不希望向服务添加任何终结点,则运行时添加默认终结点。在此示例中,由于服务的 ServiceMetadataBehavior 设置为 true,所以服务还启用了发布元数据。

(9) 打开服务主机并等待传入调用。当用户按 Enter 时,关闭服务主机。

```
svcHost.Open();
Console.WriteLine("The service is ready.");
Console.WriteLine("Press <ENTER> to terminate service.");
Console.WriteLine();Console.ReadLine();svcHost.Close();
```

(10) 生成并运行控制台应用程序。

(11) 使用 Internet Explorer 浏览服务的基址(本示例中的 http://localhost:8001/MetadataSample)并验证是否已打开元数据发布。会出现一个网页,该网页顶部显示"Simple Service"(简单服务),其下面紧接着显示"You have created a service"(已经创建服务)。如果未显示上述内容,则结果页顶部会显示消息:"Metadata publishing for this service is currently disabled"(当前禁用服务的元数据发布)。

下面的代码示例演示如何在代码中实现发布服务的元数据的基本 WCF 服务。

```
using System;using System.Runtime.Serialization;using System.ServiceModel;
using System.ServiceModel.Description;
namespace Metadata.Samples
{   [ServiceContract]
    public interface ISimpleService
    {
        [OperationContract]
        string SimpleMethod(string msg);
```

```csharp
    }
    class SimpleService : ISimpleService
    {   public string SimpleMethod(string msg)
        {
            Console.WriteLine("The caller passed in " + msg);
            return "Hello " + msg;
        }  }
    class Program
{
    static void Main(string[] args)
    {
            ServiceHost svcHost = new ServiceHost(typeof(SimpleService), new Uri("http://localhost:8001/MetadataSample"));
    try{ ServiceMetadataBehavior smb = svcHost.Description.Behaviors.Find<ServiceMetadataBehavior>();
                        if (smb == null)
                smb = new ServiceMetadataBehavior();
                smb.HttpGetEnabled = true;
                 smb.MetadataExporter.PolicyVersion = PolicyVersion.Policy15; svcHost.Description.Behaviors.Add(smb);
                svcHost.AddServiceEndpoint(
                    ServiceMetadataBehavior.MexContractName,
                    MetadataExchangeBindings.CreateMexHttpBinding(),
                    "mex"       );                                    svcHost.AddServiceEndpoint(typeof(ISimpleService), new WSHttpBinding(), "");
                //Open the service host to accept incoming calls
                svcHost.Open();
                Console.WriteLine("The service is ready.");
                Console.WriteLine("Press <ENTER> to terminate service.");
                Console.WriteLine(); Console.ReadLine();
                    svcHost.Close();     }
                catch(CommunicationException commProblem)
                {
                    Console.WriteLine("There was a communication problem. " + commProblem.Message);
                Console.Read();
            }  }  }
```

5.4 检索元数据

元数据检索是从元数据终结点(如 WS – MetadataExchange (MEX)元数据终结点或 HTTP/GET 元数据终结点)请求和检索元数据的过程。

1. 从命令行使用 Svcutil. exe 检索元数据

可以检索服务元数据通过使用 Ws – metadataexchange 或 HTTP/GET 请求 ServiceModel 元数据实用工具(Svcutil. exe)工具并传递/target：metadata 开关和地址。Svcutil. exe 在指定地址下载元数据并将文件保存到磁盘上。Svcutil. exe 在内部使用一个 System. ServiceModel. Description. MetadataExchangeClient 实例,并从配置中加载名称与作为输入传递给 Svcutil. exe 的地址方案相匹配的 IMetadataExchange 终结点配置。

2. 使用 MetadataExchangeClient 以编程方式检索元数据

Windows Communication Foundation (WCF) 可以使用标准化协议(如 WS – MetadataExchange 和 HTTP/GET 请求)检索元数据。这两种协议均受 MetadataExchangeClient 类型支持。可以通过提供元数据终结点的地址和一个可选绑定,使用 System. ServiceModel. Description. MetadataExchangeClient 类型来检索服务元数据。由 System. ServiceModel. Description. MetadataExchangeClient 实例使用的绑定可以是 MetadataExchangeBindings 静态类中的默认绑定之一、用户提供的绑定或从 IMetadataExchange 协定的终结点配置加载的绑定。System. ServiceModel. Description. MetadataExchangeClient 也可以使用 HttpWebRequest 类型来解析 HTTP URL 对元数据的引用。

默认情况下, System. ServiceModel. Description. MetadataExchangeClient 实例与单个 ChannelFactory 实例关联。通过重写 System. ServiceModel. ChannelFactory 虚拟方法,可以更改或替换由 System. ServiceModel. Description. MetadataExchangeClient 使用的 GetChannelFactory 实例。同样,通过重写 HttpWebRequest 虚拟方法,可以更改或替换由 System. ServiceModel. Description. MetadataExchangeClient 使用的 MetadataExchangeClient. GetWebRequest 实例以发出 HTTP/GET 请求。

5.4.1 使用 Svcutil. exe 下载元数据文档

可以使用 Svcutil. exe 从正在运行的服务中下载元数据并将元数据保存到本地文件。对于 HTTP 和 HTTPS URL 方案,Svcutil. exe 会尝试使用 Ws – metadataexchange 检索元数据和 XML Web 服务发现。对于所有其他 URL 架构,Svcutil. exe 仅使用 WS – MetadataExchange。

默认情况下,Svcutil.exe 使用 MetadataExchangeBindings 类中定义的绑定。若要配置用于 WS – MetadataExchange 的绑定,必须在 Svcutil.exe 的配置文件(svcutil.exe.config)中定义一个客户端终结点,使该终结点使用 IMetadataExchange 约定,并具有与元数据终结点地址的统一资源标识符(URI)架构相同的名称。

小心

当运行 Svcutil.exe 以公开两个不同的服务的服务中获取元数据协定,每个包含相同名称的操作时,Svcutil.exe 将显示"无法获取从元数据…"的错误消息。例如,如果有一个服务公开调用的服务协定具有一个操作的 ICarService Get(Car c)和同一服务还公开名为 IBookService 的具有一个 Get(Book b)操作的服务协定。若要解决此问题,请执行下列操作之一:

- √ 重命名其中的一项操作;
- √ 将 Name 设置为其他名称;
- √ 使用 Namespace 属性将其中一项操作的命名空间设置为其他命名空间。

1. 使用 Svcutil.exe 下载元数据

(1)在以下位置找到 Svcutil.exe 工具:

`C:\Program Files\Microsoft SDKs\Windows\v1.0.\bin`

(2)在命令提示符处,使用下面的格式启动该工具;

`svcutil.exe /t:metadata <url>* | <epr>`

必须指定 /t:metadata 选项才能下载元数据。否则,会生成客户端代码和配置。

(3)<url>参数指定可提供元数据的服务终结点或是联机承载的元数据文档的 URL。<epr>参数指定一个包含 Ws – addressing 的 XML 文件的路径 EndpointAddress 支持 Ws – metadataexchange 服务的终结点。

2. 示例

下面的命令从正在运行的服务中下载元数据文档。

`svcutil /t:metadata http://service/metadataEndpoint`

5.4.2 使用 MetadataResolver 动态获取绑定元数据

(1)使用元数据终结点的地址创建一个 EndpointAddress 对象。

```
EndpointAddress metaAddress
    = new EndpointAddress( new   Uri( "http://localhost:8080/SampleService/mex"));
```

(2)调用 Resolve(Type,EndpointAddress),它传入服务类型和元数据终结点地址。这将返回实现了指定协定的终结点的集合。仅从元数据导入绑定信息,不导入协定信息。改用所提供的协定。

```
ServiceEndpointCollection endpoints = MetadataResolver.Resolve(typeof
(SampleServiceClient),metaAddress);
```

（3）然后，可以遍历该服务终结点集合，以提取所需的绑定信息。下面的代码遍历终结点，创建一个服务客户端对象（该对象传入与当前终结点关联的绑定和地址），然后调用该服务上的一个方法。

```
foreach (ServiceEndpoint point in endpoints)
{    if(point! = null)
    {    using ( wcfClient = new SampleServiceClient ( point.Binding, point.Address))
    {    Console.WriteLine( wcfClient.SampleMethod("Client used the "
        + point.Address.ToString() + " address."));    }  }  }
```

5.4.3 使用 MetadataExchangeClient 检索元数据

使用 MetadataExchangeClient 类可以下载使用 WS – MetadataExchange（MEX）协议的元数据。检索到的元数据文件作为 MetadataSet 对象返回。返回的 MetadataSet 对象包含 MetadataSection 对象的集合，其中每个对象包含一个特定的元数据方言和一个标识符。您可以将返回的元数据写入文件；或者，如果返回的元数据包含 Web 服务描述语言（WSDL）文档，则可以使用 WsdlImporter 导入元数据。

接受地址的 MetadataExchangeClient 构造函数可对符合该地址的统一资源标识符（URI）方案的 MetadataExchangeBindings 静态类使用绑定。您也可以使用 MetadataExchangeClient 构造函数显式指定要使用的绑定。指定的绑定用于解析所有元数据引用。

与任何其他 Windows Communication Foundation（WCF）客户端一样，MetadataExchangeClient 类型提供一个用于使用终结点配置名称加载客户端终结点配置的构造函数。指定的终结点配置必须指定 IMetadataExchange 约定。不会加载终结点配置中的地址，因此必须使用接受地址的 GetMetadata 重载之一。当使用 EndpointAddress 实例指定元数据地址时，MetadataExchangeClient 假设该地址指向 MEX 终结点。如果将元数据地址指定为 URL，则还需要指定要使用的 MetadataExchangeClientMode(MEX 或 HTTP GET)。

重要

默认情况下，MetadataExchangeClient 解析所有引用，包括 WSDL 和 XML 架构导入和包括的内容。通过将 ResolveMetadataReferences 属性设置为 false，可以禁用此功能。使用 MaximumResolvedReferences 属性可以控制要解析的最大引用数。可以与 MaxReceivedMessageSize 属性一起对绑定使用此属性，以控制所检索的元数据的量。

使用 MetadataExchangeClient 获取元数据

（1）通过显式指定一个绑定、一个终结点配置名称或元数据的地址，创建一个新的 MetadataExchangeClient 对象。

（2）配置 MetadataExchangeClient 以适合用户的需要。例如，用户可以指定请求元数据时要使用的凭据、控制元数据引用的解析方式和设置 OperationTimeout 属性以控制元数据请求超时之前必须在多长时间内返回。

（3）通过调用 MetadataSet 方法之一获取包含检索的元数据的 GetMetadata 对象。请注意，如果您在构造 GetMetadata 时显式指定了地址，则只能使用不带参数的 MetadataExchangeClient 重载。

下面的代码示例演示如何使用 MetadataExchangeClient 下载和枚举元数据文件。

```
//Get metadata documents.
Console.WriteLine("URI of the metadata documents retreived:");
MetadataExchangeClient metaTransfer
    = new MetadataExchangeClient ( httpGetMetaAddress. Uri,
MetadataExchangeClientMode.HttpGet);
metaTransfer.ResolveMetadataReferences = true;
MetadataSet otherDocs = metaTransfer.GetMetadata();
foreach (MetadataSection doc in otherDocs.MetadataSections)
    Console.WriteLine(doc.Dialect + " : " + doc.Identifier);
```

2. 编译代码

若要编译此代码示例，必须引用 System. ServiceModel. dll 程序集并导入 System. ServiceModel. Description 命名空间。

5.5 使用元数据

服务元数据包含计算机可读的服务说明。服务元数据包括服务终结点、绑定、约定、操作和消息的说明。服务元数据有多种用途，包括自动生成使用服务的客户端、实现服务说明和动态更新客户端的绑定。

5.5.1 客户端代码

ServiceModel Metadata Utility Tool（Svcutil. exe）可生成用于生成客户端应用程序的客户端代码和客户端应用程序配置文件。

1. 概述

如果使用 Visual Studio 来生成项目的 Windows Communication Foundation（WCF）客户端类型，通常并不需要检查生成的客户端代码。如果用户并未使用为用户执行相同服务的开发环境，可以使用 Svcutil.exe 之类的工具来生成客户端代码，然后使用该代码开发客户端应用程序。

由于 Svcutil.exe 具有许多可以修改生成的类型信息的选项，因此本节并不对所有方案进行讨论。但是，以下标准任务涉及查找生成的代码：

◇ 标识服务协定接口；
◇ 标识 WCF 客户端类；
◇ 标识数据类型；
◇ 标识双工服务的回调协定；
◇ 标识帮助器服务协定通道接口。

2. 查找服务协定接口

若要查找对服务协定建模的接口，请搜索使用 System.ServiceModel.ServiceContractAttribute 属性进行标记的接口。由于存在其他属性（Attribute）和设置在该属性（Attribute）自身上的显式属性（Property），因此要通过快速读取查找该属性（Attribute）通常并不容易。请记住，服务协定接口和客户端协定接口属于两种不同的类型。下面的代码示例演示原始服务协定。

```
[ServiceContractAttribute(
  Namespace = "http://microsoft.wcf.documentation"
)]
public interface ISampleService
{
  [OperationContractAttribute]
    [FaultContractAttribute(typeof(microsoft.wcf.documentation.SampleFault))]
    string SampleMethod(string msg);
}
```

下面的代码示例演示 Svcutil.exe 生成的相同服务协定。

```
[System.ServiceModel.ServiceContractAttribute(
  Namespace = "http://microsoft.wcf.documentation"
)]
public interface ISampleService
{
```

```
[System.ServiceModel.OperationContractAttribute(
    Action = " http://microsoft.wcf.documentation/ISampleService/SampleMethod",
    ReplyAction = " http://microsoft.wcf.documentation/ISampleService/SampleMethodResponse"
)]
[System.ServiceModel.FaultContractAttribute(
    typeof(microsoft.wcf.documentation.SampleFault), Action = " http://microsoft.wcf.documentation/ISampleService/SampleMethodSampleFaultFault"    )]
string SampleMethod(string msg);}
```

可以使用生成的服务协定接口和 System.ServiceModel.ChannelFactory 类一起创建 WCF 通道对象,使用该通道对象可调用服务操作。

3. 查找 WCF 客户端类

若要查找实现用户要使用的服务协定的 WCF 客户端类,请搜索 System.ServiceModel.ClientBase < TChannel > 的扩展名,其中类型参数为用户先前查找到的且扩展该接口的服务协定接口。下面的代码示例演示类型为 ClientBase < TChannel > 的 ISampleService 类。

```
[System.CodeDom.Compiler.GeneratedCodeAttribute("System.ServiceModel", "3.0.0.0")]
public partial class SampleServiceClient : System.ServiceModel.ClientBase < ISampleService > , ISampleService
{
    public SampleServiceClient()
    { }
public SampleServiceClient ( string endpointConfigurationName ): base(endpointConfigurationName)
    { }
     public SampleServiceClient ( string endpointConfigurationName, string remoteAddress) : base(endpointConfigurationName, remoteAddress)
    { }
     public SampleServiceClient ( string endpointConfigurationName, System.ServiceModel.EndpointAddress remoteAddress)
        :
            base(endpointConfigurationName, remoteAddress)
    { }
    public SampleServiceClient(System.ServiceModel.Channels.Binding binding, System.ServiceModel.EndpointAddress remoteAddress)
```

```
            :base(binding, remoteAddress)
        {    }
    public string SampleMethod(string msg)
        {        return base.Channel.SampleMethod(msg);      }}
```

可以使用此 WCF 客户端类,方法是创建它的一个新实例并调用它实现的方法。这些方法调用服务操作,可设计和配置该服务操作以进行交互。

备注

SvcUtil.exe 在生成 WCF 客户端类时会将一个 DebuggerStepThroughAttribute 添加到该客户端类,以防止调试器逐步调试该 WCF 客户端类。

4. 查找数据类型

若要在生成的代码中查找数据类型,最简单的方法就是标识协定中指定的类型名称并搜索该类型声明的代码。例如,下面的协定指定 SampleMethod 可以返回 microsoft.wcf.documentation.SampleFault 类型的 SOAP 错误。

```
[System.ServiceModel.OperationContractAttribute(
   Action = "http://microsoft.wcf.documentation/ISampleService/SampleMethod",
    ReplyAction = " http://microsoft. wcf. documentation/ISampleService/SampleMethodResponse"
)]
[System.ServiceModel.FaultContractAttribute(
   typeof(microsoft.wcf.documentation.SampleFault),
    Action = " http://microsoft. wcf. documentation/ISampleService/SampleMethodSampleFaultFault"
)]
string SampleMethod(string msg);
```

搜索 SampleFault 查找下面的类型声明。

```
[assembly: System.Runtime.Serialization.ContractNamespaceAttribute(
   "http://microsoft.wcf.documentation",
   ClrNamespace = "microsoft.wcf.documentation")
]
namespace microsoft.wcf.documentation
{
    using System.Runtime.Serialization;
     [System.CodeDom.Compiler.GeneratedCodeAttribute(" System.Runtime.Serialization","3.0.0.0")]
     [System.Runtime.Serialization.DataContractAttribute()]
```

```
        public partial class SampleFault : object, System.Runtime.
Serialization.IExtensibleDataObject
        {
                private System.Runtime.Serialization.ExtensionDataObject
extensionDataField;
            private string FaultMessageField;
                        public System.Runtime.Serialization.
ExtensionDataObject ExtensionData
        {get{return this.extensionDataField;}
            set
            {this.extensionDataField = value;}
        }
        [System.Runtime.Serialization.DataMemberAttribute()]
        public string FaultMessage
        {get{return this.FaultMessageField;}
            set
            {this.FaultMessageField = value;}}}
```

在本示例中,数据类型为客户端上的特定异常(一个 FaultException < TDetail >,其中详细信息类型为 microsoft.wcf.documentation.SampleFault)引发的详细信息类型。

5. 查找双工服务的回调协定

如果要查找协定接口为其 ServiceContractAttribute.CallbackContract 属性指定一个值的服务协定,则该协定指定一个双工协定。双工协定要求客户端应用程序创建一个回调类,该类实现回调协定并将此类的一个实例传递给 System.ServiceModel.DuplexClientBase < TChannel > 或用来与服务进行通信的 System.ServiceModel.DuplexChannelFactory < TChannel >。

下面的协定指定一个 SampleDuplexHelloCallback 类型的回调协定。

```
[System.ServiceModel.ServiceContractAttribute(
  Namespace = "http://microsoft.wcf.documentation",
  ConfigurationName = "SampleDuplexHello",
  CallbackContract = typeof(SampleDuplexHelloCallback),
  SessionMode = System.ServiceModel.SessionMode.Required)]
public interface SampleDuplexHello
{[System.ServiceModel.OperationContractAttribute(
  IsOneWay = true,
  Action = "http://microsoft.wcf.documentation/SampleDuplexHello/Hello"
  )] void Hello(string greeting);  }
```

搜索该回调协定定位下面的接口,客户端应用程序必须实现该接口。

```
[System.CodeDom.Compiler.GeneratedCodeAttribute("System.ServiceModel","3.0.0.0")]
    public interface SampleDuplexHelloCallback
{[System.ServiceModel.OperationContractAttribute(
        IsOneWay = true,
        Action = "http://microsoft.wcf.documentation/SampleDuplexHello/Reply"
    )]void Reply(string responseToGreeting);}
```

6. 查找服务协定通道接口

当 ChannelFactory 类与服务协定接口一起使用时，必须强制转换到 System.ServiceModel.IClientChannel 接口以显式打开、关闭或中止通道。为便于操作，Svcutil.exe 工具还生成一个帮助器接口，该接口可同时实现服务协定接口和 IClientChannel，无须进行强制转换就能够与客户端通道基础结构进行交互。下面的代码演示实现上述服务协定的帮助器客户端通道的定义。

```
[System.CodeDom.Compiler.GeneratedCodeAttribute("System.ServiceModel","3.0.0.0")]
    public interface ISampleServiceChannel : ISampleService, System.ServiceModel.IClientChannel
    {}
```

5.5.2 检索元数据并实现兼容服务

通常，设计和实现服务并不是由同一个人完成的。在交互操作应用程序很重要的环境中，可以用 Web 服务描述语言（WSDL）设计或描述协定，而且开发人员必须实现一个与所提供的协定相兼容的服务。此外，双工协定还需要调用方实现一个回调协定。

在这些情况下，必须使用 ServiceModel 元数据实用工具（Svcutil.exe）（或等效的工具）以托管语言的方式生成服务协定下的接口协定。通常情况下，Svcutil.exe 用于获取一个服务协定，该服务协定与通道工厂或 WCF 客户端以及设置正确的绑定和地址的客户端配置文件一起使用。若要使用生成的配置文件，则必须将其更改到服务配置文件中。用户可能还需要修改服务协定。

1. 检索数据并实现兼容服务

（1）对元数据文件或元数据终结点使用 Svcutil.exe 生成代码文件。

（2）搜索输出代码文件中包含相关接口的部分（以防存在多个接口），此接口是用 System.ServiceModel.ServiceContractAttribute 属性标记的。下面的代码演示由 Svcutil.exe 生成两个接口。第一个（ISampleService）是服务协定接口，实现它可创建兼容服务。第二个（ISampleServiceChannel）是帮助器接口，客户端使用它可同时扩展服务协定接口和 System.

ServiceModel.IClientChannel,且该接口可用于客户端应用程序。

```
[System.CodeDom.Compiler.GeneratedCodeAttribute("System.ServiceModel", "3.0.
0.0")]
[System.ServiceModel.ServiceContractAttribute(
   Namespace = "http://microsoft.wcf.documentation"
)]
public interface ISampleService
{ [System.ServiceModel.OperationContractAttribute(
         Action = " http://microsoft. wcf. documentation/ISampleService/
SampleMethod",
         ReplyAction = " http://microsoft. wcf. documentation/ISampleService/
SampleMethodResponse" )]
      [System.ServiceModel.FaultContractAttribute(
        typeof(microsoft.wcf.documentation.SampleFault),
         Action = " http://microsoft. wcf. documentation/ISampleService/
SampleMethodSampleFaultFault"
      )] string SampleMethod(string msg);
}
[System.CodeDom.Compiler.GeneratedCodeAttribute("System.ServiceModel", "3.0.
0.0")]
public   interface   ISampleServiceChannel   :   ISampleService,   System.
ServiceModel.IClientChannel
{}
```

（3）如果 WSDL 未指定所有操作的答复操作,则生成的操作协定可能会将 ReplyAction 属性设置为通配符（）。移除该属性设置,否则,当实现服务协定元数据时,将不能为这些操作导出元数据。

（4）实现类上的接口并承载服务。

（5）在客户端的配置文件 ServiceModel 元数据实用工具（Svcutil.exe）生成时,更改 <客户端> 的 <服务> 配置节。（有关生成的客户端应用程序配置文件的示例,请参见下面的示例部分。）

（6）在 <服务> 配置部分中,创建 name 属性中 <服务> 配置节,用于服务实现。

（7）将服务的 name 属性设置为服务实现的配置名称。

（8）将使用实现的服务协定的终结点配置元素添加到服务配置部分。

2. 示例

下面的代码通过演示运行元数据文件 Svcutil.exe 而生成大部分代码文件。

下面的代码演示:
◇ 实现时符合协定要求的服务协定接口(ISampleService);
◇ 客户端所使用的帮助器接口,可用于同时扩展服务协定接口和 System. ServiceModel.IClientChannel,并可用于客户端应用程序(ISampleServiceChannel);
◇ 扩展 System.ServiceModel.ClientBase < TChannel > 的帮助器类,可用于客户端应用程序(SampleServiceClient);
◇ 从服务生成的配置文件;
◇ 简单的 ISampleService 服务实现;
◇ 客户端配置文件到服务器端版本的转换。

```
//------------------------------------------------------------
// <auto-generated>
//     This code was generated by a tool.
//     Runtime Version:2.0.50727.42
//
//     Changes to this file may cause incorrect behavior and will be lost if
//     the code is regenerated.
// </auto-generated>
//------------------------------------------------------------


[assembly: System.Runtime.Serialization.ContractNamespaceAttribute("http://microsoft.wcf.documentation", ClrNamespace = "microsoft.wcf.documentation")]

namespace microsoft.wcf.documentation
{
    using System.Runtime.Serialization;

    [System.CodeDom.Compiler.GeneratedCodeAttribute("System.Runtime.Serialization", "3.0.0.0")]
    [System.Runtime.Serialization.DataContractAttribute()]
    public partial class SampleFault : object, System.Runtime.Serialization.IExtensibleDataObject
    {
        private System.Runtime.Serialization.ExtensionDataObject extensionDataField;
        private string FaultMessageField;
        public System.Runtime.Serialization.ExtensionDataObject ExtensionData
        {
            get { return this.extensionDataField; }
```

```csharp
            set { this.extensionDataField = value; } }
    [System.Runtime.Serialization.DataMemberAttribute()]
    public string FaultMessage
            { get { return this.FaultMessageField; }
            set { this.FaultMessageField = value; } } }
[System.CodeDom.Compiler.GeneratedCodeAttribute("System.ServiceModel", "3.0.0.0")]
[System.ServiceModel.ServiceContractAttribute(
    Namespace = "http://microsoft.wcf.documentation"
)]
public interface ISampleService
    { [System.ServiceModel.OperationContractAttribute(
        Action = " http://microsoft. wcf. documentation/ISampleService/SampleMethod",
        ReplyAction = " http://microsoft. wcf. documentation/ISampleService/SampleMethodResponse" )]
    [System.ServiceModel.FaultContractAttribute(
        typeof(microsoft.wcf.documentation.SampleFault),
        Action = " http://microsoft. wcf. documentation/ISampleService/SampleMethodSampleFaultFault" )]
    string SampleMethod(string msg);
    }
[System.CodeDom.Compiler.GeneratedCodeAttribute("System.ServiceModel", "3.0.0.0")]
public interface ISampleServiceChannel : ISampleService, System.ServiceModel.IClientChannel
    { }
[System.CodeDom.Compiler.GeneratedCodeAttribute("System.ServiceModel", "3.0.0.0")]
public partial class SampleServiceClient : System.ServiceModel.ClientBase<ISampleService>, ISampleService
    { public SampleServiceClient()
        { }
    public SampleServiceClient(string endpointConfigurationName) :
            base(endpointConfigurationName)
        { }
        public SampleServiceClient ( string endpointConfigurationName, string
```

```
remoteAddress) :
            base(endpointConfigurationName, remoteAddress)
    { }
    public SampleServiceClient( string endpointConfigurationName, System.ServiceModel.EndpointAddress remoteAddress) :
            base(endpointConfigurationName, remoteAddress)
    { }
    public SampleServiceClient(System.ServiceModel.Channels.Binding binding, System.ServiceModel.EndpointAddress remoteAddress) :
            base(binding, remoteAddress)
    { }
    public string SampleMethod(string msg)
    {
        return base.Channel.SampleMethod(msg);
    }
}
```

XML 文件:

```xml
<configuration>
    <system.serviceModel>
        <bindings>
            <basicHttpBinding>
                <binding name="BasicHttpBinding_ISampleService" closeTimeout="00:01:00"
                    openTimeout="00:01:00" receiveTimeout="00:10:00" sendTimeout="00:01:00"
                    allowCookies="false" bypassProxyOnLocal="false" hostNameComparisonMode="StrongWildcard"
                    maxBufferSize="65536" maxBufferPoolSize="524288" maxReceivedMessageSize="65536"
                    messageEncoding="Text" textEncoding="utf-8" transferMode="Buffered"
                    useDefaultWebProxy="true">
                    <readerQuotas maxDepth="32" maxStringContentLength="8192" maxArrayLength="16384"
                        maxBytesPerRead="4096" maxNameTableCharCount="16384" />
                    <security mode="None">
                        <transport clientCredentialType="None"
```

```
proxyCredentialType = "None"
                                  realm = "" />
                        < message clientCredentialType = " UserName "
algorithmSuite = "Default" />
                        < /security >
                    < /binding >
                < /basicHttpBinding >
            < /bindings >
            < client >
                < endpoint address = "http://localhost:8080/SampleService" binding
= "basicHttpBinding"
                       bindingConfiguration = " BasicHttpBinding _ ISampleService "
contract = "ISampleService"
                       name = "BasicHttpBinding_ISampleService" />
            < /client >
        < /system.serviceModel >
    < /configuration >
    // Implement the service. This is a very simple service.
    class SampleService : ISampleService
    {   public string SampleMethod( string msg)
        {
         Console.WriteLine( "The caller said: \"{0}\"", msg);
         return "The service greets you: " + msg;
        } }
```

XML 文件:

```
< configuration >
  < system.serviceModel >
    < bindings >
      < basicHttpBinding >
        < binding name = "BasicHttpBinding_ISampleService" closeTimeout = "00:
01:00"
            openTimeout = "00:01:00" receiveTimeout = "00:10:00" sendTimeout = "00:
01:00"
                    allowCookies = " false " bypassProxyOnLocal = " false "
hostNameComparisonMode = "StrongWildcard"
                    maxBufferSize = " 65536 " maxBufferPoolSize = " 524288 "
maxReceivedMessageSize = "65536"
```

```xml
                messageEncoding="Text" textEncoding="utf-8" transferMode="Buffered"
                useDefaultWebProxy="true">
                    <readerQuotas maxDepth="32" maxStringContentLength="8192" maxArrayLength="16384"
                        maxBytesPerRead="4096" maxNameTableCharCount="16384" />
                    <security mode="None">
                        <transport clientCredentialType="None" proxyCredentialType="None"
                            realm="" />
                        <message clientCredentialType="UserName" algorithmSuite="Default" />
                    </security>
                </binding>
            </basicHttpBinding>
        </bindings>
        <services>
            <service
                name="Microsoft.WCF.Documentation.SampleService">
                <endpoint address="http://localhost:8080/SampleService" binding="basicHttpBinding"
                    bindingConfiguration="BasicHttpBinding_ISampleService" contract="Microsoft.WCF.Documentation.ISampleService"
                 />
            </service>
        </services>
    </system.serviceModel>
</configuration>
```

第6章 客户端

6.1 WCF 客户端体系结构

应用程序使用 Windows Communication Foundation（WCF）客户端对象来调用服务操作。

1. 概述

服务模型运行时创建 WCF 客户端,这些客户端由以下各项构成:

自动生成的服务协定客户端实现,该实现可以将应用程序代码中的调用转换为传出消息,将响应消息转换为输出参数并返回应用程序可以检索的值。

控制接口（System. ServiceModel. IClientChannel）的实现,它可将各个接口组合在一起,并提供对控制功能（特别是用于关闭客户端会话和释放通道的能力）的访问。

根据已有绑定的配置属性生成对应的客户端通道。

应用程序可以根据需要通过 System. ServiceModel. ChannelFactory 或通过创建由 Svcutil. exe 生成的 ClientBase 派生类的实例来创建这样的客户端。这些预先生成的客户端类可以封装并委托给由 ChannelFactory 动态构造的客户端通道实现。因此,客户端通道和生成这些客户端通道的客户端工厂是此处讨论的焦点。

2. 客户端对象和客户端通道

WCF 客户端的基接口是 System. ServiceModel. IClientChannel 接口,该接口公开核心客户端功能以及 System. ServiceModel. ICommunicationObject 的基本通信对象功能、System. ServiceModel. IContextChannel 的上下文功能和 System. ServiceModel. IExtensibleObject < T > 的可扩展行为。

但 IClientChannel 接口并不定义服务协定本身。这些协定通过服务协定接口进行声明（通常使用 Svcutil. exe 这样的工具从服务元数据生成）。WCF 客户端类型同时扩展 IClientChannel 和目标服务协定接口,以使应用程序能够直接调用操作,同时还可以访问客户端运行时功能。创建一个 WCF 客户端可以向 WCFSystem. ServiceModel. ChannelFactory 对象提供必要信息,以便创建可以连接到配置的服务终结点并与此终结点交互的运行时。

如前所述,这两个 WCF 客户端类型必须进行配置才能使用。最简单的 WCF 客户端类型是从 ClientBase < TChannel > 中派生的对象(如果服务协定是双工协定,则为从 DuplexClientBase < TChannel > 中派生的对象)。通过使用以编程方式进行配置的构造函数,或者通过使用配置文件,然后直接调用该配置文件以调用服务操作,可以创建这些类型。

第二个类型是在运行时从对 CreateChannel 方法的调用中生成的。通常涉及严格控制通信详细信息的应用程序使用此客户端类型,调用客户端通道对象,因为基础客户端运行时和通道系统相比,它可以启用更直接的交互。

3. 通道工厂

负责创建可支持客户端调用的基础运行时的类为 System.ServiceModel.ChannelFactory < TChannel > 类。WCF 客户端对象和 WCF 客户端通道对象都使用 ChannelFactory < TChannel > 对象来创建实例;ClientBase < TChannel > 派生的客户端对象封装通道工厂的处理,但在有些情况下,完全可以直接使用通道工厂。对此,常见的情况是从现有工厂重复创建新的客户端通道。如果要使用客户端对象,可以通过调用 WCF 属性从 ClientBase < TChannel >.ChannelFactory 客户端对象中获取基础通道工厂。

应该记住的是,在调用 ChannelFactory < TChannel >.CreateChannel 之前,通道工厂会为所提供的配置创建客户端通道的新实例。一旦调用 CreateChannel(或 ClientBase < TChannel >.Open、ClientBase < TChannel >.CreateChannel 或对 WCF 客户端对象执行的任何操作)后,不能修改通道工厂和期待获取到不同服务实例的通道,即使只是更改目标终结点地址也是如此。如果想用不同配置创建客户端对象或客户端通道,则必须首先创建新的通道工厂。

以下说明 WCF 客户端通道对象的创建和使用。

(1)创建新的 WCF 客户端通道对象

为了说明客户端通道的用法,假设已生成以下服务协定。

```
[System.ServiceModel.ServiceContractAttribute(
  Namespace = "http://microsoft.wcf.documentation"
)]
public interface ISampleService
{
    [System.ServiceModel.OperationContractAttribute(
        Action = " http://microsoft.wcf.documentation/ISampleService/SampleMethod",
        ReplyAction = " http://microsoft.wcf.documentation/ISampleService/SampleMethodResponse"
    )]
    [System.ServiceModel.FaultContractAttribute(
```

```
       typeof(microsoft.wcf.documentation.SampleFault),
           Action  =  " http://microsoft. wcf. documentation/ISampleService/
SampleMethodSampleFaultFault"
       )]
       string SampleMethod(string msg);
   }
```

若要连接到 ISampleService 服务,请直接与通道工厂(ChannelFactory < TChannel >)一起使用生成的协定接口。一旦为特定协定创建并配置通道工厂后,即可以调用 CreateChannel 方法以返回可用于与 ISampleService 服务通信的客户端通道对象。

当 ChannelFactory < TChannel > 类与服务协定接口一起使用时,必须强制转换到 IClientChannel 接口以显式打开、关闭或中止通道。为便于操作,Svcutil.exe 工具还生成一个帮助器接口,该接口可同时实现服务协定接口和 IClientChannel,以使用户无须进行强制转换就能够与客户端通道基础结构进行交互。下面的代码演示实现上述服务协定的帮助器客户端通道的定义。

```
   [System.CodeDom.Compiler.GeneratedCodeAttribute("System.ServiceModel", "3.0.
0.0")]
   public    interface    ISampleServiceChannel    :    ISampleService, System.
ServiceModel.IClientChannel
   {}
```

(2)创建新的 WCF 客户端通道对象

若要使用客户端通道连接到 ISampleService 服务,请直接与通道工厂一起使用生成的协定接口(或帮助器版本),将协定接口的类型作为参数进行传递。一旦为特定协定创建并配置通道工厂后,即可以调用 ChannelFactory < TChannel > .CreateChannel 方法以返回可用于与 ISampleService 服务通信的客户端通道对象。

客户端通道对象在创建后可实现 IClientChannel 和协定接口。因此,可以直接使用这些对象来调用操作,与支持该协定的服务进行交互。

使用客户端对象和客户端通道对象之间的区别仅仅在于其易于使用和开发人员的喜好。许多习惯于使用类和对象的开发人员喜欢使用 WCF 客户端对象,而不喜欢使用 WCF 客户端通道。

6.2 WCF 客户端访问服务

6.2.1 概述

本主题描述与以下内容相关的行为和问题：
◇ 通道和会话生存期；
◇ 处理异常；
◇ 了解阻塞问题；
◇ 以交互方式初始化通道；
◇ 通道和会话生存期。

Windows Communication Foundation（WCF）应用程序包含两个类别的通道：数据报通道和会话通道。

数据报通道是在其中所有消息都是不相关的通道。使用数据报通道时，如果输入或输出操作失败，下一个操作通常不会受到影响，并且同一个通道可以重用。因此，数据报通道通常不会出错。

会话通道，是与另一个终结点连接的通道。某一端会话中的消息总是与另一端的同一会话相关联。另外，会话的两个参与者必须商定，只有彼此的对话要求得到满足，才能认为该会话是成功的。如果他们无法商定，则会话通道可能会出错。

备注

试图显式检测出错的会话通道通常是没有用处的，因为用户何时得到通知取决于会话实现。例如，因为 System.ServiceModel.NetTcpBinding(禁用了可靠会话)表现了 TCP 连接会话的状态，所以，如果在服务或客户端上侦听 ICommunicationObject.Faulted 事件，则在出现网络故障时，可能会很快得到通知。但是，可靠会话（由启用了 System.ServiceModel.Channels.ReliableSessionBindingElement 的绑定建立）旨在防止服务受到小型网络故障的影响。如果可以在一段合理的时间内重新建立会话，则同一绑定（为可靠会话而配置）可能不会出错，除非中断持续了一段较长的时间。

默认情况下，大多数由系统提供的绑定（它们向应用程序层公开通道）都使用会话，但 System.ServiceModel.BasicHttpBinding 不使用会话。

（1）正确使用会话

会话提供了一种了解整个消息交换是否已完成以及会话双方是否都认为交换成功的方式。建议让调用应用程序在一个 try 块内打开、使用和关闭通道。如果会话通道打开，然后调用了一次 ICommunicationObject.Close 方法,并且该调用成功返回,则会话是成功的。在此情况下,成功意味着绑定所指定的所有传递保证都得到满足,并且另一方在调用 ICommunicationObject.Abort 之前没有对通道调用 Close。

下一节提供了此客户端方法的一个示例。

(2)处理异常

在客户端应用程序中处理异常是非常简单的。如果在一个 try 块内打开、使用并关闭通道,则除非引发了异常,否则对话都是成功的。通常情况下,如果引发了异常,则会中止对话。

备注

建议不要使用 using 语句(在 using 中为 Visual Basic)。这是因为 using 语句的末尾会引发异常,这些异常会屏蔽其他异常。

下面的代码示例演示使用 try/catch 块而不是 using 语句的推荐客户端模式。

```
using System;using System.ServiceModel;using System.ServiceModel.Channels;
using Microsoft.WCF.Documentation;
public class Client
{ public static void Main()
  { SampleServiceClient wcfClient = new SampleServiceClient();
    try { Console.WriteLine("Enter the greeting to send: ");
      string greeting = Console.ReadLine();
      Console.WriteLine("The service responded: " + wcfClient.SampleMethod(greeting));
      Console.WriteLine("Press ENTER to exit:"); Console.ReadLine();
      wcfClient.Close(); Console.WriteLine("Done!"); }
    catch (TimeoutException timeProblem)
    {
      Console.WriteLine("The service operation timed out. " + timeProblem.Message);
      Console.ReadLine(); wcfClient.Abort(); }
    catch (FaultException<GreetingFault> greetingFault)
    { Console.WriteLine(greetingFault.Detail.Message);
      Console.ReadLine(); wcfClient.Abort(); }
    catch (FaultException unknownFault)
    {
      Console.WriteLine("An unknown exception was received. " + unknownFault.Message);
```

```
            Console.ReadLine();
            wcfClient.Abort();          }
        catch (CommunicationException commProblem)
        {
            Console.WriteLine("There was a communication problem. " + commProblem.
Message + commProblem.StackTrace);
            Console.ReadLine();
            wcfClient.Abort();
        }    }    }
```

备注

　　检查 ICommunicationObject.State 属性的值是一个争用条件,建议不要使用此检查来确定是重用还是关闭通道。

　　即使在关闭数据报通道时出现异常,这些通道也从来不会出错。另外,使用安全对话但未能通过身份验证的非双工客户端通常会引发 System.ServiceModel.Security.MessageSecurityException。但是,与此不同的是,如果使用安全对话的双工客户端未能通过身份验证,该客户端会收到 System.TimeoutException。

　　预期异常描述预期的异常,并显示如何处理它们。

　　(3) 客户端阻塞和性能

　　当应用程序同步调用请求 – 答复操作时,客户端会阻塞,直到接收到返回值或引发异常(例如 System.TimeoutException)为止。此行为与本地行为类似。当应用程序同步调用 WCF 客户端对象或通道上的操作时,客户端将直到通道层可以向网络写入数据或引发异常时才返回。虽然单向消息交换模式(通过标记操作来指定,即将 OperationContractAttribute.IsOneWay 设置为 true)可以使某些客户端更快做出响应,但是根据绑定和已经发送的消息的性质,单向操作也可能阻塞。单向操作仅与消息交换有关。

　　无论使用何种消息交换模式,大的数据块都会降低客户端的处理速度。

　　如果应用程序在操作完成过程中必须做更多的工作,则应在 WCF 客户端实现的服务协定接口上创建一个异步方法对。若要执行此操作的最简单方法是使用/async 切换 ServiceModel 元数据实用工具(Svcutil.exe)。

　　(4) 使用户可以动态选择凭据

　　IInteractiveChannelInitializer 接口使应用程序可以显示一个用户界面,用户可以使用该界面选择凭据,以便用来在超时计时器启动之前创建通道。

　　应用程序开发人员可以通过两种方式利用一个插入的 IInteractiveChannelInitializer。客户端应用程序可以调用 ClientBase < TChannel >.DisplayInitializationUI 或 IClientChannel.DisplayInitializationUI(或异步版本)在打开通道之前(显式方法)或调用的第一个操作(隐

式方法)。

如果使用隐式方法,则应用程序必须调用 ClientBase < TChannel > 或 IClientChannel 扩展上的第一个操作。如果它调用除第一个操作以外的任何操作,则将引发异常。

如果使用显式方法,应用程序必须按顺序执行下面的操作:

① 调用 ClientBase < TChannel >. DisplayInitializationUI 或 IClientChannel.DisplayInitializationUI(或异步版本);

② 当初始值设定项已返回时,针对 Open 对象或从 IClientChannel 属性返回的 IClientChannel 对象来调用 ClientBase < TChannel >. InnerChannel 方法;

③ 调用操作。

建议采用显式方法,由可投入实际生产运行的应用程序来控制用户界面过程。

使用隐式方法的应用程序调用用户界面初始值设定项,但是如果应用程序的用户没有在绑定的发送超时期限内做出响应,则当用户界面返回时,将引发异常。

6.2.2 使用单向和请求-答复协定访问 WCF 服务

下面的过程描述了如何访问一个 Windows Communication Foundation (WCF) 服务,该服务定义一个单向协定和一个请求-答复协定,并且未使用双工通信模式。

1. 定义服务

(1) 声明服务协定。要成为单向的操作必须在 IsOneWay 中将 true 设置为 OperationContractAttribute。下面的代码声明具有 IOneWayCalculator、Add、Subtract 和 Multiply 的单向操作的 Divide 协定。它还定义称为 SayHello 的请求响应操作。

```
[ServiceContract(Namespace = "http:/Microsoft.ServiceModel.Samples")]
public interface IOneWayCalculator
{ [OperationContract(IsOneWay = true)]  void Add(double n1, double n2);
  [OperationContract(IsOneWay = true)]  void Subtract(double n1, double n2);
  [OperationContract(IsOneWay = true)]  void Multiply(double n1, double n2);
  [OperationContract(IsOneWay = true)]  void Divide(double n1, double n2);
    [OperationContract]  string SayHello(string name);  }
```

(2) 实现服务协定。下面的代码实现 IOnewayCalculator 接口。

```
[ ServiceBehavior ( ConcurrencyMode = ConcurrencyMode. Multiple,
InstanceContextMode = InstanceContextMode.PerCall)]
public class CalculatorService : IOneWayCalculator
{     public void Add(double n1, double n2)
    {
        double result = n1 + n2;
```

```csharp
        Console.WriteLine("Add({0},{1}) = {2}", n1, n2, result);
    }
    public void Subtract(double n1, double n2)
    {
       double result = n1 - n2;
       Console.WriteLine("Subtract({0},{1}) = {2}", n1, n2, result);
    }
    public void Multiply(double n1, double n2)
    {
       double result = n1 * n2;
       Console.WriteLine("Multiply({0},{1}) = {2}", n1, n2, result);
    }
    public void Divide(double n1, double n2)
    {   double result = n1 /n2;
       Console.WriteLine("Divide({0},{1}) = {2}", n1, n2, result);
    }
    public string SayHello(string name)
    {  Console.WriteLine("SayHello({0})", name);
       return "Hello " + name;
    }  }
```

2. 在控制台应用程序中承载服务

下面的代码演示如何承载服务。

```csharp
public static void Main()
{Uri baseAddress = new Uri ( " http://localhost:8000/servicemodelsamples/service");
    using (ServiceHost serviceHost = new ServiceHost(typeof(CalculatorService), baseAddress))
    { serviceHost. AddServiceEndpoint ( typeof ( IOneWayCalculator ), new WSHttpBinding(), "");
    ServiceMetadataBehavior smb = ( ServiceMetadataBehavior ) serviceHost. Description.Behaviors.Find < ServiceMetadataBehavior > ();
       if (smb == null)
       {  smb = new ServiceMetadataBehavior();
           smb.HttpGetEnabled = true;
           serviceHost.Description.Behaviors.Add(smb);
       }
```

```
serviceHost.Open();
Console.WriteLine("The service is ready.");
Console.WriteLine("Press <ENTER> to terminate service.");
Console.WriteLine(); Console.ReadLine();    }  }
```

3. 访问服务

运行 ServiceModel 元数据实用工具（Svcutil.exe）使用元数据交换终结点地址创建使用下面的命令行的服务的客户端类：Svcutil http://localhost:8000/Service ServiceModel 元数据实用工具（Svcutil.exe）生成一组接口和类，如下面的代码所示。

```
[System.CodeDom.Compiler.GeneratedCodeAttribute("System.ServiceModel", "3.0.0.0")]
[System.ServiceModel.ServiceContractAttribute(Namespace = "http://Microsoft.ServiceModel.Samples", ConfigurationName = "IOneWayCalculator")]
public interface IOneWayCalculator
{
    [System.ServiceModel.OperationContractAttribute(IsOneWay = true, Action = "http://Microsoft.ServiceModel.Samples/IOneWayCalculator/Add")]
    void Add(double n1, double n2);
    [System.ServiceModel.OperationContractAttribute(IsOneWay = true, Action = "http://Microsoft.ServiceModel.Samples/IOneWayCalculator/Subtract")]
    void Subtract(double n1, double n2);
    [System.ServiceModel.OperationContractAttribute(IsOneWay = true, Action = "http://Microsoft.ServiceModel.Samples/IOneWayCalculator/Multiply")]
    void Multiply(double n1, double n2);
    [System.ServiceModel.OperationContractAttribute(IsOneWay = true, Action = "http://Microsoft.ServiceModel.Samples/IOneWayCalculator/Divide")]
    void Divide(double n1, double n2);
    [System.ServiceModel.OperationContractAttribute(Action = "http://Microsoft.ServiceModel.Samples/IOneWayCalculator/SayHello", ReplyAction = "http://Microsoft.ServiceModel.Samples/IOneWayCalculator/SayHelloResponse")]
    string SayHello(string name);
}
[System.CodeDom.Compiler.GeneratedCodeAttribute("System.ServiceModel", "3.0.0.0")]
public interface IOneWayCalculatorChannel : IOneWayCalculator, System.ServiceModel.IClientChannel
{  }
```

```csharp
[System.Diagnostics.DebuggerStepThroughAttribute()]
[System.CodeDom.Compiler.GeneratedCodeAttribute("System.ServiceModel", "3.0.0.0")]
public partial class OneWayCalculatorClient : System.ServiceModel.ClientBase<IOneWayCalculator>, IOneWayCalculator
{
    public OneWayCalculatorClient()
    {   }
    public OneWayCalculatorClient(string endpointConfigurationName) :
            base(endpointConfigurationName)
    {   }
    public OneWayCalculatorClient(string endpointConfigurationName, string remoteAddress) :
            base(endpointConfigurationName, remoteAddress)
    {   }
    public OneWayCalculatorClient(string endpointConfigurationName, System.ServiceModel.EndpointAddress remoteAddress) :
            base(endpointConfigurationName, remoteAddress)
    {   }
    public OneWayCalculatorClient(System.ServiceModel.Channels.Binding binding, System.ServiceModel.EndpointAddress remoteAddress) :
            base(binding, remoteAddress)
    {   }
    public void Add(double n1, double n2)
    {       base.Channel.Add(n1, n2);       }
    public void Subtract(double n1, double n2)
    {       base.Channel.Subtract(n1, n2);       }
    public void Multiply(double n1, double n2)
    {       base.Channel.Multiply(n1, n2);       }
    public void Divide(double n1, double n2)
    {       base.Channel.Divide(n1, n2);       }
    public string SayHello(string name)
    {       return base.Channel.SayHello(name);       }  }
```

请注意，在 IOneWayCalculator 接口中，单向服务操作已将 IsOneWay 属性设置为 true，请求－答复服务操作已将属性设置为默认值 false。此外，请注意 OneWayCalculatorClient 类。这是将用于调用服务的类。

第6章 客 户 端

（1）创建客户端对象。
```
//Create a client
WSHttpBinding binding = new WSHttpBinding();
EndpointAddress epAddress = new EndpointAddress("http://localhost:8000/servicemodelsamples/service");
OneWayCalculatorClient client = new OneWayCalculatorClient(binding, epAddress);
```
（2）调用服务操作。
```
//Call the Add service operation.
double value1 = 100.00D;
double value2 = 15.99D;
client.Add(value1, value2);
Console.WriteLine("Add({0},{1})", value1, value2);
//Call the Subtract service operation.
value1 = 145.00D;
value2 = 76.54D;
client.Subtract(value1, value2);
Console.WriteLine("Subtract({0},{1})", value1, value2);
//Call the Multiply service operation.
value1 = 9.00D;
value2 = 81.25D;
client.Multiply(value1, value2);
Console.WriteLine("Multiply({0},{1})", value1, value2);
//Call the Divide service operation.
value1 = 22.00D;
value2 = 7.00D;
client.Divide(value1, value2);
Console.WriteLine("Divide({0},{1})", value1, value2);
//Call the SayHello service operation
string name = "World";
string response = client.SayHello(name);
Console.WriteLine("SayHello([0])", name);
Console.WriteLine("SayHello() returned: " + response);
```
（3）关闭客户端,以便关闭连接并清理资源。
```
//Closing the client gracefully closes the connection and cleans up resources
client.Close();
```
示例

下面列出了本主题中使用的完整代码。

```csharp
//Service.cs
using System; using System.Configuration;using System.ServiceModel;
using System.ServiceModel.Description;
namespace Microsoft.ServiceModel.Samples
{
    //Define a service contract.
    [ServiceContract(Namespace = "http://Microsoft.ServiceModel.Samples")]
    public interface IOneWayCalculator
    {
        [OperationContract(IsOneWay = true)] void Add(double n1, double n2);
        [OperationContract(IsOneWay = true)] void Subtract(double n1, double n2);
        [OperationContract(IsOneWay = true)] void Multiply(double n1, double n2);
        [OperationContract(IsOneWay = true)] void Divide(double n1, double n2);
        [OperationContract] string SayHello(string name);
    }
    [ServiceBehavior(ConcurrencyMode = ConcurrencyMode.Multiple, InstanceContextMode = InstanceContextMode.PerCall)]
    public class CalculatorService : IOneWayCalculator
    {
        public void Add(double n1, double n2)
        {
            double result = n1 + n2;
            Console.WriteLine("Add({0},{1}) = {2}", n1, n2, result);
        }
        public void Subtract(double n1, double n2)
        {
            double result = n1 - n2;
            Console.WriteLine("Subtract({0},{1}) = {2}", n1, n2, result);
        }
        public void Multiply(double n1, double n2)
        {
            double result = n1 * n2;
            Console.WriteLine("Multiply({0},{1}) = {2}", n1, n2, result);
        }
        public void Divide(double n1, double n2)
        {
            double result = n1 / n2;
            Console.WriteLine("Divide({0},{1}) = {2}", n1, n2, result);
        }
        public string SayHello(string name)
        {
            Console.WriteLine("SayHello({0})", name);
```

```csharp
            return "Hello " + name;
        }
        //Host the service within this EXE console application.
        public static void Main()
        {
Uri baseAddress = new Uri(" http://localhost:8000/servicemodelsamples/service");
using (ServiceHost serviceHost = new ServiceHost(typeof(CalculatorService), baseAddress)) { serviceHost.AddServiceEndpoint(typeof(IOneWayCalculator), new WSHttpBinding(),"");
ServiceMetadataBehavior smb = (ServiceMetadataBehavior) serviceHost.Description.Behaviors.Find<ServiceMetadataBehavior>();
            if (smb == null)
            {
                smb = new ServiceMetadataBehavior();
                smb.HttpGetEnabled = true;
                serviceHost.Description.Behaviors.Add(smb);
            }
serviceHost.Open(); Console.WriteLine("The service is ready.");
Console.WriteLine("Press <ENTER> to terminate service.");
Console.WriteLine(); Console.ReadLine(); } } }
using System; using System.ServiceModel;
namespace Microsoft.ServiceModel.Samples
{   class Client
    {
        static void Main()
        { WSHttpBinding binding = new WSHttpBinding();
EndpointAddress epAddress = new EndpointAddress(" http://localhost:8000/servicemodelsamples/service "); OneWayCalculatorClient client = new OneWayCalculatorClient(binding, epAddress); double value1 = 100.00D; double value2 = 15.99D; client.Add(value1, value2);
Console.WriteLine("Add({0},{1})", value1, value2);
            //Call the Subtract service operation.
            value1 = 145.00D;  value2 = 76.54D;
            client.Subtract(value1, value2);
            Console.WriteLine("Subtract({0},{1})", value1, value2);
            //Call the Multiply service operation.
```

```
            value1 = 9.00D;    value2 = 81.25D;
            client.Multiply(value1, value2);
            Console.WriteLine("Multiply({0},{1})", value1, value2);
            //Call the Divide service operation.
value1 = 22.00D; value2 = 7.00D; client.Divide(value1, value2);
Console.WriteLine("Divide({0},{1})", value1, value2);
            //Call the SayHello service operation
            string name = "World";
            string response = client.SayHello(name);
            Console.WriteLine("SayHello([0])", name);
            Console.WriteLine("SayHello() returned: " + response);
            client.Close();  }  }  }
```

6.2.3 使用双工协定访问服务

Windows Communication Foundation（WCF）的一个功能是可以创建使用双工消息传递模式的服务。此模式允许服务通过回调与客户端进行通信。本节讲述在实现回调接口的客户端类中创建 WCF 客户端的步骤。

双向绑定向服务公开客户端的 IP 地址。客户端应使用安全来确保仅连接到自己信任的服务。

访问双工服务

（1）创建包含两个接口的服务。第一个接口用于服务，第二个接口用于回调。

```
[ ServiceContract ( Namespace = " http://Microsoft.ServiceModel.Samples ",
SessionMode = SessionMode.Required,
            CallbackContract = typeof(ICalculatorDuplexCallback))]
public interface ICalculatorDuplex
{
    [OperationContract(IsOneWay = true)]
    void Clear();
    [OperationContract(IsOneWay = true)]
    void AddTo(double n);
    [OperationContract(IsOneWay = true)]
    void SubtractFrom(double n);
    [OperationContract(IsOneWay = true)]
    void MultiplyBy(double n);
    [OperationContract(IsOneWay = true)]
    void DivideBy(double n);
```

```
}
public interface ICalculatorDuplexCallback
{
    [OperationContract(IsOneWay = true)]
    void Equals(double result);
    [OperationContract(IsOneWay = true)]
    void Equation(string eqn);
}
```

（2）运行服务。

（3）使用 ServiceModel 元数据实用工具（Svcutil.exe）来为客户端生成协定（接口）。

（4）在客户端类中实现回调接口，如下面的示例所示。

```
public class CallbackHandler : ICalculatorDuplexCallback
{
    public void Result(double result)
    {
        Console.WriteLine("Result ({0})", result);
    }
    public void Equation(string equation)
    {
        Console.WriteLine("Equation({0})", equation);
    }
}
```

（5）创建 InstanceContext 类的一个实例。构造函数需要客户端类的一个实例。

```
InstanceContext site = new InstanceContext(new CallbackHandler());
```

（6）使用需要 WCF 对象的构造函数创建 InstanceContext 客户端的一个实例。该构造函数的第二个参数是配置文件中找到的终结点的名称。

```
CalculatorDuplexClient wcfClient =
new CalculatorDuplexClient(site, "default")
```

（7）根据需要调用 WCF 客户端的方法。

下面的代码演示如何创建一个访问双工协定的客户端类。

```
using System;
using System.ServiceModel;
using System.ServiceModel.Channels;

// Define class that implements the callback interface of duplex
// contract.
```

```csharp
public class CallbackHandler : ICalculatorDuplexCallback
{
    public void Result(double result)
    {
        Console.WriteLine("Result({0})", result);
    }
    public void Equation(string equation)
    {
        Console.WriteLine("Equation({0})", equation);
    }
}
public class Client
{
    public static void Main()
    {CalculatorDuplexClient wcfClient
        = new CalculatorDuplexClient(new InstanceContext(new CallbackHandler()));
        try
        { double value = 100.00D;
          wcfClient.AddTo(value);
          value = 50.00D;
          wcfClient.SubtractFrom(value);
          //Call the MultiplyBy service operation.
          value = 17.65D;
          wcfClient.MultiplyBy(value);
          //Call the DivideBy service operation.
          value = 2.00D;
          wcfClient.DivideBy(value);
          //Complete equation.
          wcfClient.Clear();
          //Wait for callback messages to complete before
          //closing.
          System.Threading.Thread.Sleep(5000);
          //Close the WCF client.
          wcfClient.Close();
          Console.WriteLine("Done!");
        }
        catch (TimeoutException timeProblem)
```

```
            }
            Console.WriteLine("The service operation timed out. " + timeProblem.
Message);
            wcfClient.Abort();
            Console.Read();
        }
        catch (CommunicationException commProblem)
        {
            Console.WriteLine("There was a communication problem. " + commProblem.
Message);
            wcfClient.Abort();
            Console.Read();
        }    }
```

6.2.4 使用 ChannelFactory

ChannelFactory<TChannel> 泛型类用于某些高级方案中,这些方案要求创建可用于创建多个通道的通道工厂。

创建和使用 ChannelFactory 类

生成并运行一个 Windows Communication Foundation(WCF)服务。

使用 ServiceModel 元数据实用工具(Svcutil.exe)生成客户端的协定(接口)。

在客户端代码中,使用 ChannelFactory<TChannel> 类创建多个终结点侦听器。

示例如下:

```
using System;using System.ServiceModel;
[ServiceContract()]
interface IMath
{
   [OperationContract()]
    double Add(double A, double B);}
public class Math : IMath
{    public double Add(double A, double B)
    {        return A + B;    }}
public sealed class Test
{    static void Main()    {    }
    public void Run()
    {    BasicHttpBinding myBinding = new BasicHttpBinding();
        EndpointAddress myEndpoint = new EndpointAddress("http://localhost/
```

```
MathService/Epl");
        ChannelFactory < IMath > myChannelFactory = new ChannelFactory < IMath
>(myBinding, myEndpoint);
    IMath wcfClient1 = myChannelFactory.CreateChannel();
    double s = wcfClient1.Add(3, 39);
    Console.WriteLine(s.ToString());
    ((IClientChannel)wcfClient1).Close();
        IMath wcfClient2 = myChannelFactory.CreateChannel();
        s = wcfClient2.Add(15, 27);
        Console.WriteLine(s.ToString());
    ((IClientChannel)wcfClient2).Close();
    myChannelFactory.Close();
        }}
```

6.2.5　以异步方式调用 WCF 服务操作

本节介绍客户端如何以异步方式访问服务操作。本节中的服务可实现 ICalculator 接口。通过使用事件驱动的异步调用模型,客户端可以对此接口异步调用操作。

备注

使用 ChannelFactory < TChannel > 时,不支持事件驱动的异步调用模型。

过程

以异步方式调用 WCF 服务操作

(1) 运行 ServiceModel 元数据实用工具(Svcutil. exe)两个工具/async 和/tcv:Version35 一起命令选项,如下面的命令中所示。

svcutil /n:http://Microsoft.ServiceModel.Samples, Microsoft.ServiceModel. Samples http://localhost:8000/servicemodelsamples/service/mex /a /tcv:Version35

除了同步操作和基于委托的标准异步操作之外,此操作还会生成 WCF 客户端类,其中包含以下内容:

◇　两个 < operationName > Async 用于基于事件的异步调用方法的操作。

例如:

```
public void AddAsync(double n1, double n2)
{
    this.AddAsync(n1, n2, null);
}
public void AddAsync(double n1, double n2, object userState)
{
```

```
        if ((this.onBeginAddDelegate = = null))
        {
                this. onBeginAddDelegate = new BeginOperationDelegate ( this.
OnBeginAdd);
        }
        if ((this.onEndAddDelegate = = null))
        {
            this.onEndAddDelegate = new EndOperationDelegate(this.OnEndAdd);
        }
        if ((this.onAddCompletedDelegate = = null))
        {
                this. onAddCompletedDelegate = new System. Threading.
SendOrPostCallback(this.OnAddCompleted);
        }
        base.InvokeAsync(this.onBeginAddDelegate, new object[] {
                n1,
                    n2}, this. onEndAddDelegate, this. onAddCompletedDelegate,
userState);
    }
```

◇ 窗体的操作已完成事件 < operationName > Completed 用于基于事件的异步调用方法。

例如：

```
public event System.EventHandler<AddCompletedEventArgs> AddCompleted;
```

◇ System. EventArgs 每个操作的类型（窗体的 < operationName > CompletedEventArgs）用于基于事件的异步调用方法。

例如：

```
[System.Diagnostics.DebuggerStepThroughAttribute()]
[System.CodeDom.Compiler.GeneratedCodeAttribute("System.ServiceModel", "3.0.0.0")]
    public partial class AddCompletedEventArgs : System. ComponentModel.
AsyncCompletedEventArgs
    {
        private object[] results;

            public AddCompletedEventArgs ( object [ ] results, System. Exception
exception, bool cancelled, object userState) :
```

```
            base(exception, cancelled, userState)
    {    this.results = results;        }

    public double Result
    {
        get        {
            base.RaiseExceptionIfNecessary();
            return ((double)(this.results[0]));
        }
    }
}
```

（2）在调用应用程序中，创建一个要在异步操作完成时调用的回调方法，如下面的示例代码所示。

```
//Asynchronous callbacks for displaying results.
static void AddCallback(object sender, AddCompletedEventArgs e)
{
    Console.WriteLine("Add Result: {0}", e.Result);
}
```

（3）在调用该操作之前，使用一个新的泛型 System.EventHandler < TEventArgs > 类型的 < operationName > EventArgs 将（在上一步中创建）的处理程序方法添加到 < operationName > Completed 事件。然后调用 < operationName > Async 方法。例如：

```
//AddAsync
double value1 = 100.00D;
double value2 = 15.99D;
client.AddCompleted + = new EventHandler < AddCompletedEventArgs > (AddCallback);
client.AddAsync(value1, value2);
Console.WriteLine("Add({0},{1})", value1, value2);
```

示例

备注

基于事件的异步模型设计准则规定，如果返回了多个值，则一个值会作为 Result 属性返回，其他值会作为 EventArgs 对象上的属性返回。因此产生的结果之一是，如果客户端使用基于事件的异步命令选项导入元数据，且该操作返回多个值，则默认的 EventArgs 对象返回一个值作为 Result 属性，返回的其余值是 EventArgs 对象的属性。如果要将消息对象作为 Result 属性来接收并要使返回的值作为该对象上的属性，请使用 /messageContract 命令选项。这会生成一个签名,该签名会将响应消息作为 Result 对象上的 EventArgs 属性返回。然后,所有内部返回值就都是响应消息对象的属性了。

```csharp
using System;
namespace Microsoft.ServiceModel.Samples
{
    //The service contract is defined in generatedClient.cs, generated from the service by the svcutil tool.
    class Client
    { static void Main()
        {
            Console.WriteLine("Press <ENTER> to terminate client once the output is displayed.");
            Console.WriteLine();
            CalculatorClient client = new CalculatorClient();
            double value1 = 100.00D;
            double value2 = 15.99D;
            client.AddCompleted += new EventHandler<AddCompletedEventArgs>(AddCallback);
            client.AddAsync(value1, value2);
            Console.WriteLine("Add({0},{1})", value1, value2);
            value1 = 145.00D; value2 = 76.54D;
            client.SubtractCompleted += new EventHandler<SubtractCompletedEventArgs>(SubtractCallback);
            client.SubtractAsync(value1, value2);
            Console.WriteLine("Subtract({0},{1})", value1, value2);
            value1 = 9.00D; value2 = 81.25D;
            double result = client.Multiply(value1, value2);
            Console.WriteLine("Multiply({0},{1}) = {2}", value1, value2, result);
            value1 = 22.00D; value2 = 7.00D;
            result = client.Divide(value1, value2);
            Console.WriteLine("Divide({0},{1}) = {2}", value1, value2, result);
            Console.ReadLine(); client.Close();          }
        static void AddCallback(object sender, AddCompletedEventArgs e)
        {    Console.WriteLine("Add Result: {0}", e.Result);          }
        static void SubtractCallback(object sender, SubtractCompletedEventArgs e)
```

            { Console.WriteLine("Subtract Result: {0}", e.Result); }   }   }

6.2.6 使用通道工厂以异步方式调用操作

本节介绍客户端如何在基于 ChannelFactory < TChannel > 的客户端应用程序时以异步方式访问服务操作。当使用 System. ServiceModel. ClientBase < TChannel > 对象调用服务时，可以使用事件驱动的异步调用模型。

本节中的服务可实现 ICalculator 接口。客户端可以在此接口上以异步方式调用操作，这意味着像 Add 这样的操作将拆分为两个方法：BeginAdd 和 EndAdd。前一个方法启动调用，而后一个方法在操作完成时检索结果。

过程

以异步方式调用 WCF 服务操作：

（1）运行 ServiceModel 元数据实用工具（Svcutil. exe）工具/async 选项如下面的命令中所示。

svcutil /n：http://Microsoft. ServiceModel. Samples, Microsoft. ServiceModel. Samples http://localhost:8000/servicemodelsamples/service/mex /a

这将生成该操作的服务协定的异步客户端版本。

（2）创建一个异步操作完成时将要调用的回调函数，如下面的示例代码所示。

```
static void AddCallback(IAsyncResult ar)
{
    double result = ((CalculatorClient)ar.AsyncState).EndAdd(ar);
    Console.WriteLine("Add Result: {0}", result);
}
```

（3）若要以异步方式访问服务操作，请创建客户端、调用 Begin[Operation]（例如 BeginAdd），并指定一个回调函数，如下面的示例代码所示。

```
ChannelFactory<ICalculatorChannel> factory = new ChannelFactory<ICalculatorChannel>();
ICalculatorChannel channelClient = factory.CreateChannel();
double value1 = 100.00D;
double value2 = 15.99D;
IAsyncResult arAdd = channelClient.BeginAdd(value1, value2, AddCallback, channelClient);
Console.WriteLine("Add({0},{1})", value1, value2);
```

执行该回调函数时，客户端将调用 End < operation >（例如 EndAdd）以检索结果。

示例

上面过程中的客户端代码使用的服务可实现 ICalculator 接口，如下面的代码所示。在服务端,协定的 Add 和 Subtract 操作由 Windows Communication Foundation（WCF）运行库以同步方式调用,即使前面的客户端步骤是在客户端以异步方式调用的也是如此。在服务端,Multiply 和 Divide 操作用于在服务端以异步方式调用服务,即使客户端以同步方式调用这两个操作也是如此。此示例将 AsyncPattern 属性设置为 true。此属性设置与．NET Framework 异步模式的实现一起,可让运行库以异步方式调用该操作。

```
[ServiceContract(Namespace = "http://Microsoft.ServiceModel.Samples")]
public interface ICalculator
{   [OperationContract]
    double Add(double n1, double n2);
    [OperationContract]
    double Subtract(double n1, double n2);
    //Multiply involves some file I/O so we'll make it Async.
    [OperationContract(AsyncPattern = true)]
    IAsyncResult BeginMultiply(double n1, double n2, AsyncCallback callback, object state);
    double EndMultiply(IAsyncResult ar);
    //Divide involves some file I/O so we'll make it Async.
    [OperationContract(AsyncPattern = true)]
    IAsyncResult BeginDivide(double n1, double n2, AsyncCallback callback, object state);
    double EndDivide(IAsyncResult ar);}
```

6.2.7 创建通道工厂并用它创建和管理通道

通过 DuplexChannelFactory < TChannel > 类可以创建和管理不同类型的双工通道,客户端可以使用这些通道在服务终结点之间发送和接收消息。

示例

下面的代码演示如何创建通道工厂并用它来创建和管理通道。

```
InstanceContext site = new InstanceContext(new ChatApp());
using (DuplexChannelFactory < IChatChannel > cf =
    new DuplexChannelFactory < IChatChannel > (site,"ChatEndpoint"))
{
    X509Certificate2 issuer = GetCertificate(
    StoreName.CertificateAuthority,
    StoreLocation.CurrentUser, "CN = " + issuerName,
    X509FindType.FindBySubjectDistinguishedName);
    cf.Credentials.Peer.Certificate =
```

```
GetCertificate(StoreName.My,
StoreLocation.CurrentUser,
"CN = " + member,
X509FindType.FindBySubjectDistinguishedName);
cf.Credentials.Peer.PeerAuthentication.CertificateValidationMode =
X509CertificateValidationMode.Custom;
cf.Credentials.Peer.PeerAuthentication.CustomCertificateValidator =
new IssuerBasedValidator();
using (IChatChannel participant = cf.CreateChannel())
{   IOnlineStatus ostat = participant.GetProperty<IOnlineStatus>();
ostat.Online += new EventHandler(OnOnline);
ostat.Offline += new EventHandler(OnOffline);
Console.WriteLine("{0} is ready", member);
Console.WriteLine("Press <ENTER> to send the chat message.");
participant.Join(member);
Console.ReadLine();
participant.Chat(member, "Hi there - I am chatting");
Console.WriteLine("Press <ENTER> to terminate this instance of chat.");
Console.ReadLine();
participant.Leave(member);
}}
```

6.3　WCF 客户端配置

可以使用 Windows Communication Foundation（WCF）客户端配置来指定客户端终结点的地址（Address）、绑定（Binding）、行为（Behavior）和协定（Contract），即客户端终结点的 ABC 属性来连接服务终结点。<客户端>元素具有<终结点>，其属性用于配置终结点的 ABC 属性的元素。这些属性将在本主题的"配置终结点"一节中讨论。

<终结点>元素还包含<元数据>元素，用于指定元数据的导入和导出，一个<headers>元素（包含自定义地址标头集合）和一个<identity>元素（通过该元素，终结点可与其他终结点交换信息，实现身份验证）。<标头>和<标识>元素属于 EndpointAddress 进行讨论地址主题。本节的"配置元数据"子节中提供了一些主题链接，这些主题对元数据扩展的使用进行说明。

1. 配置终结点

通过客户端配置，客户端可以指定一个或多个终结点，每个终结点都有自己的名称、地址和协定,并且每个终结点都引用客户端配置中要用于配置该终结点的<bindings>和<behaviors>元素。客户端配置文件应命名为"App.config",因为这是 WCF 运行库所约定的名称。下面的示例演示一个客户端配置文件。

```xml
<?xml version="1.0" encoding="utf-8"?>
<configuration>
  <system.serviceModel>
    <client>
      <endpoint
        name="endpoint1"
        address="http://localhost/ServiceModelSamples/service.svc"
        binding="wsHttpBinding"
        bindingConfiguration="WSHttpBinding_IHello"
        behaviorConfiguration="IHello_Behavior"
        contract="IHello">
        <metadata>
          <wsdlImporters>
            <extension
                              type="Microsoft.ServiceModel.Samples.WsdlDocumentationImporter,WsdlDocumentation"/>
          </wsdlImporters>
        </metadata>
        <identity>
          <servicePrincipalName value="host/localhost" />
        </identity>
      </endpoint>
//Add another endpoint by adding another <endpoint> element.
      <endpoint
        name="endpoint2">
        //Configure another endpoint here.
      </endpoint>
    </client>
//The bindings section references by the bindingConfiguration endpoint attribute.
    <bindings>
      <wsHttpBinding>
        <binding name="WSHttpBinding_IHello"
            bypassProxyOnLocal="false"
```

```
                    hostNameComparisonMode = "StrongWildcard" >
                <readerQuotas maxDepth = "32" />
                <reliableSession ordered = "true"
                        enabled = "false" />
                <security mode = "Message" >
                //Security settings go here.
                </security>
            </binding>
            <binding name = "Another Binding"
            //Configure this binding here.
            </binding>
              </wsHttpBinding>
            </bindings>
            <behaviors>
              <endpointBehaviors>
                <behavior name = "IHello_Behavior" >
                    <clientVia />
                </behavior>
              </endpointBehaviors>
            </behaviors>
        </system.serviceModel>
    </configuration>
```

可选的 name 属性唯一地标识了给定协定的终结点。它由 ChannelFactory < TChannel > 或 ClientBase < TChannel > 用于指定客户端配置中的哪个终结点是目标终结点,必须在创建到服务的通道时加载。通配符终结点配置名称"＊"可用,并且指示 ApplyConfiguration 方法,如果文件中正好有一个终结点配置,就应加载该终结点配置,否则引发异常。如果省略此属性,则将对应的终结点用作与指定协定类型相关联的默认终结点。name 属性的默认值是一个空字符串,它与任何其他名称一样进行匹配。

每个终结点都必须具有一个与之关联的地址,用于查找和标识终结点。address 属性可用来指定提供终结点位置的 URL。但是,通过创建统一资源标识符（URI）,也可以在代码中指定服务终结点的地址,使用 ServiceHost 方法之一可以将该地址添加到 AddServiceEndpoint。正如简介所述,< headers > 和 < identity > 元素是 EndpointAddress 的组成部分,终结点地址中也对此进行了论述。

binding 属性指示终结点在连接到服务时期望使用的绑定类型。该类型必须具有一个已注册的配置节,才能加以引用。在前面的示例中,这是 < wsHttpBinding > 节,它指示终结点使用 WSHttpBinding。实际上,终结点可以使用某个给定类型的多个绑定。其中每个具有其自己<绑定>在（binding）类型元素的元素。bindingconfiguration 属性用于区分相同类型的绑定。其值与 name 属性匹配的<绑定>元素。

behaviorConfiguration 属性用于指定该<行为>的 <endpointBehaviors>终结点应使用。其值与<behavior>元素的 name 属性匹配。

contract 属性指定终结点公开哪个协定。此值对应于 ConfigurationName 的 ServiceContractAttribute。默认值为实现相应服务的类的完整类型名。

2. 配置元数据

<元数据>元素用于指定用于注册元数据的设置导入扩展。

第7章 承 载

7.1 IIS 承载

若要承载 Windows Communication Foundation（WCF）服务，一种选择是在 Internet 信息服务（IIS）应用程序内部承载。此承载模型与 ASP.NET 和 ASP.NET Web 服务（ASMX）使用的模型类似。

1. IIS 的版本

可以在下列操作系统上的以下 IIS 版本上承载 WCF：

Windows XP SP2 上的 IIS 5.1。此环境对于设计和开发 IIS 承载的应用程序非常有用，这些应用程序稍后将部署在 Windows Server 2003 等服务器操作系统上。

Windows Server 2003 上的 IIS 6.0 提供了一种高级进程模型，这种模型可提供更好的可伸缩性、可靠性和应用程序隔离。此环境适合对以独占方式使用 HTTP 通信的 WCF 服务进行成品部署。

Windows Vista 和 Windows Server 2008 上的 IIS 7.0。IIS 7.0 提供了与 IIS 6.0 相同的高级进程模型，但它使用 Windows 进程激活服务（WAS）来允许通过 HTTP 之外的协议进行激活和网络通信。WCF 支持的任何网络协议（包括 HTTP、net.tcp、net.pipe 和 net.msmq）进行通信的 WCF 服务。

2. IIS 承载的好处

在 IIS 中承载 WCF 服务具有以下几个好处：

可像处理其他任何类型的 IIS 应用程序（包括 WCF 应用程序和 ASMX）一样，部署和管理 IIS 中承载的 ASP.NET 服务。

IIS 提供进程激活、运行状况管理和回收功能以提高承载的应用程序的可靠性。

像 ASP.NET 一样，WCF 中承载的 ASP.NET 服务可以利用 ASP.NET 共享宿主模型。在此模型中，多个应用程序驻留在一个公共辅助进程中以提高服务器密度和可伸缩性。

IIS 中承载的 WCF 服务与 ASP.NET 2.0 使用相同的动态编译模型，该模型简化了承载的服务的开发和部署。

当决定在 IIS 中承载 WCF 服务时,一定要记住 IIS 5.1 和 IIS 6.0 仅限于 HTTP 通信。

3. 部署 IIS 承载的 WCF 服务

开发和部署 IIS 承载的 WCF 服务由以下任务组成:
(1)请确保正确安装和注册 IIS、ASP.NET、WCF 和 WCF HTTP 激活组件。
(2)创建新的 IIS 应用程序,或重新使用现有的 ASP.NET 应用程序。
(3)为 WCF 服务创建 .svc 文件。
(4)将服务实现部署到 IIS 应用程序。
(5)配置 WCF 服务。

4. WCF 服务和 ASP.NET

WCF 服务既可以与 ASP.NET 并行承载,也可以在 ASP.NET 兼容模式中承载。在该模式下,服务可以充分利用 ASP.NET Web 应用程序平台提供的功能。

7.1.1 IIS 部署承载 WCF 服务

开发和部署承载于 Internet 信息服务(IIS)中的 Windows Communication Foundation(WCF)服务包括以下任务:
◇ 请确保正确安装和注册 IIS、ASP.NET、WCF 和 WCF 激活组件;
◇ 创建新的 IIS 应用程序,或重新使用现有的 ASP.NET 应用程序;
◇ 为 WCF 服务创建 .svc 文件;
◇ 将服务实现部署到 IIS 应用程序;
◇ 配置 WCF 服务。

(1)确保已正确安装和注册 IIS、ASP.NET 和 WCF

必须同时安装 WCF、IIS 和 ASP.NET,承载于 IIS 中的 WCF 服务才能正常工作。安装 WCF(作为 WinFX 的一部分)、ASP.NET 和 IIS 的过程因所使用的操作系统版本不同而不同。

如果计算机上已存在 IIS,则在安装 WinFX 的过程中自动向 IIS 注册 WCF。如果在安装 WinFX 后安装 IIS,则需要执行其他步骤向 IIS 和 WCF 注册 ASP.NET。根据操作系统,可以按如下所述执行此操作:

◇ Windows XP、Windows 7 和 Windows Server 2003:使用 ServiceModel 注册工具(ServiceModelReg.exe)工具注册 WCF 与 IIS:若要使用此工具,键入 ServiceModelReg.exe /i /x 中 Visual Studio 命令提示符。打开此命令提示符,方法是单击"开始"按钮、选择"所有程序","Microsoft Visual Studio 2012","Visual Studio 工具"和"Visual Studio 命令提示符"。

◇ Windows Vista:安装 WinFX 的 Windows Communication Foundation Activation

Components 子组件。为此,请在控制面板中,单击添加或删除程序然后添加/删除 Windows 组件。这将激活"Windows 组件向导"。

最后,必须确认已将 ASP.NET 配置为使用 .NET Framework 版本 4。通过运行带 -i 选项的 ASPNET_Regiis 工具执行此操作。

(2) 创建新的 IIS 应用程序或重新使用现有的 ASP.NET 应用程序

承载于 IIS 中的 WCF 服务必须驻留在 IIS 应用程序内。可以创建一个新的 IIS 应用程序专门承载 WCF 服务。或者,也可以将 WCF 服务部署到现有应用程序,该应用程序已经承载 ASP.NET 2.0 内容(如 .aspx 页和 ASP.NET Web 服务[ASMX])。

请注意,IIS 6.0 和更高版本定期重新启动独立的面向对象编程应用程序。默认值为 1740 分钟。支持的最大值为 71 582 分钟。可以禁用此重新启动。

(3) 为 WCF 服务创建 .svc 文件

承载于 IIS 中的 WCF 服务在 IIS 应用程序内表示为特殊内容文件(.svc 文件)。此模型与在 IIS 应用程序内将 ASMX 页表示为 .asmx 文件的方式类似。.svc 文件包含 WCF 特定的处理指令(@ ServiceHost),该指令允许 WCF 承载基础结构激活所承载的服务以响应传入消息。.svc 文件的最常见语法如下语句所示。

```
<% @ ServiceHost Service = " MyNamespace.MyServiceImplementationTypeName " % >
```

它由 @ ServiceHost 指令和单个属性 Service 组成。Service 属性的值是服务实现的公共语言运行库(CLR)类型名称。使用此指令与使用以下代码创建服务主机基本等效。

```
new ServiceHost( typeof( MyNamespace.MyServiceImplementationTypeName ) );
```

也可以执行其他承载配置,如创建服务的基址列表。也可以使用自定义 ServiceHostFactory 扩展指令以用于自定义承载解决方案。承载 WCF 服务的 IIS 应用程序不负责管理 ServiceHost 实例的创建和生存期。收到 .svc 文件的第一个请求时,托管 WCF 承载基础结构动态创建必需的 ServiceHost 实例。在代码显式关闭该实例之前或回收应用程序时,不释放该实例。

(4) 将服务实现部署到 IIS 应用程序

承载于 IIS 中的 WCF 服务与 ASP.NET 2.0 使用相同的动态编译模型。就像在 ASP.NET 中那样,可以在各种位置通过几种方式部署承载于 IIS 中的 WCF 服务的实现代码,如下所示:

◇ 作为全局程序集缓存(GAC)或应用程序的 \bin 目录中的预编译 .dll 文件。在部署类库的新版本后,才更新预编译的二进制文件。

◇ 作为位于应用程序的 \App_Code 目录中的未编译源文件。处理应用程序的第一个请求时,动态需要位于此目录中的源文件。对 \App_Code 目录中文件进行的任何更改都导致在收到下一个请求时回收和重新编译整个应用程序。

第 7 章 承 载

◇ 作为直接放置在 .svc 文件中的未编译代码。实现代码也可以按内联方式位于服务的 .svc 文件之后，@ServiceHost 指令。对内联代码进行的任何更改导致在收到下一个请求时回收和重新编译应用程序。

(5) 配置 WCF 服务

承载于 IIS 中的 WCF 服务将其配置存储在应用程序 Web.config 文件中。承载于 IIS 中的服务与承载于 IIS 外部的 WCF 服务使用相同的配置元素和语法。但是，下面的约束对 IIS 承载环境是唯一的：

◇ 承载于 IIS 中的服务的基址；
◇ 应用程序承载 WCF 在 IIS 外部的服务可以控制通过将一组基址 Uri 对传递这些服务的基址 ServiceHost 构造函数或通过提供 <主机> 中服务的配置元素。承载于 IIS 中的服务无法控制其基址；承载于 IIS 中的服务的基址是其 .svc 文件的地址。

承载于 IIS 中的服务的终结点地址：

承载于 IIS 中时，任何终结点地址始终被认为相对于表示服务的 .svc 文件的地址。例如，如果 WCF 服务的基址是包含以下终结点配置的 http://localhost/Application1/MyService.svc。

< endpoint address = "anotherEndpoint" ... />

这提供了一个可以在"http://localhost/Application1/MyService.svc/anotherEndpoint"上访问的终结点。

同样，将空字符串用作相对地址的终结点配置元素提供了一个可以在 http://localhost/Application1/MyService.svc（它是基址）上访问的终结点。

< endpoint address = "" ... />

对于承载于 IIS 中的服务的终结点，必须始终使用相对终结点地址。如果终结点地址未指向承载公开终结点的服务的 IIS 应用程序，则提供完全限定的终结点地址（例如，http://localhost/MyService.svc）可能导致在部署服务时出错。对所承载的服务使用相对终结点地址避免了这些潜在冲突。

(6) 可用传输

承载于 IIS 5.1 和 WCF 中的 IIS 6.0 服务被限制为使用基于 HTTP 的通信。在这些 IIS 平台上，将所承载的服务配置为使用非 HTTP 绑定会导致服务激活期间出错。对于 IIS 7.0，支持的传输包括 HTTP、Net.TCP、Net.Pipe、Net.MSMQ，以及用于与现有 MSMQ 应用程序向后兼容的 msmq.formatname。

(7) HTTP 传输安全

承载于 IIS 中的 WCF 服务可以使用 HTTP 传输安全（例如 HTTPS 和 HTTP 身份验证方案，如基本、摘要式和 Windows 集成身份验证），前提是包含该服务的 IIS 虚拟目录支持这些设置。所承载终结点的绑定上的 HTTP 传输安全设置必须与包含它的 IIS 虚拟目录上的传输安全设置匹配。

例如,配置为使用 HTTP 摘要式身份验证的 WCF 终结点必须驻留在也配置为允许 HTTP 摘要式身份验证的 IIS 虚拟目录中。IIS 设置和 WCF 终结点设置的不匹配组合会导致服务激活期间出错。

7.1.2 IIS 承载 WCF 服务方法

本主题概述了创建 Internet 信息服务（IIS）中承载的 Windows Communication Foundation（WCF）服务所需的基本步骤。本节假设用户熟悉 IIS 且了解如何使用 IIS 管理工具创建和管理 IIS 应用程序。有关以下内容的详细信息请参阅 IIS Internet Information Services。在 IIS 环境中运行的 WCF 服务可充分利用 IIS 功能,如进程回收、空闲时关闭、进程运行状况监视和基于消息的激活。此宿主选项要求正确配置 IIS,但不需要编写任何承载代码作为应用程序的一部分。只可以将 IIS 宿主与 HTTP 传输协议一起使用。

1. 创建由 IIS 承载的服务

（1）确认 IIS 已经安装并在计算机上运行。有关以下内容的详细信息安装和配置 IIS 请参阅安装和配置 IIS 7.0。

（2）为应用程序文件创建一个称为"IISHostedCalcService"的新文件夹,确保 ASP.NET 有权访问该文件夹的内容,并使用 IIS 管理工具创建一个物理上位于此应用程序目录中的新 IIS 应用程序。当为应用程序目录创建别名时,请使用"IISHostedCalc"。

（3）在应用程序目录中创建一个名为"service.svc"的新文件。通过添加以下编辑此文件@ServiceHost 元素。

```
<%@ ServiceHost language = c# Debug = "true" Service = "Microsoft.ServiceModel.Samples.CalculatorService"%>
```

（4）在应用程序目录中创建 App_Code 子目录。

（5）在 App_Code 子目录中创建名为 Service.cs 的代码文件。

（6）将下面的 using 语句添加到 Service.cs 文件的最前面。

```
using System;
using System.ServiceModel;
```

（7）将下面的命名空间声明添加到 using 语句后面。

```
namespace Microsoft.ServiceModel.Samples
{ }
```

（8）定义命名空间声明中的服务协定,如下面的代码所示。

```
[ServiceContract]
public interface ICalculator
{
```

```
    [OperationContract]
    double Add(double n1, double n2);
    [OperationContract]
    double Subtract(double n1, double n2);
    [OperationContract]
    double Multiply(double n1, double n2);
    [OperationContract]
    double Divide(double n1, double n2);
}
```

(9)在服务协定定义后实现服务协定,如下面的代码所示。

```
public class CalculatorService : ICalculator
{
    public double Add(double n1, double n2)
    {
        return n1 + n2;
    }
    public double Subtract(double n1, double n2)
    {
        return n1 - n2;
    }
    public double Multiply(double n1, double n2)
    {
        return n1 * n2;
    }
    public double Divide(double n1, double n2)
    {
        return n1 /n2;
    }
}
```

(10)在应用程序目录中创建一个名为"Web.config"的文件,并将下面的配置代码添加到该文件中。在运行时,WCF 基础结构使用这些信息来构造客户端应用程序可与其通信的终结点。

```
<?xml version = "1.0" encoding = "utf-8"?>
<configuration>
  <system.serviceModel>
    <services>
      <service name = "Microsoft.ServiceModel.Samples.CalculatorService">
```

```xml
        <!-- This endpoint is exposed at the base address provided by host: http://localhost/servicemodelsamples/service.svc -->
        <endpoint address = ""
                  binding = "wsHttpBinding"
                  contract = "Microsoft.ServiceModel.Samples.ICalculator" />
        <!-- The mex endpoint is explosed at http://localhost/servicemodelsamples/service.svc/mex -->
        <endpoint address = "mex"
                  binding = "mexHttpBinding"
                  contract = "IMetadataExchange" />
      </service>
    </services>
  </system.serviceModel>
</configuration>
```

此示例显式指定配置文件中的终结点。如果不希望向服务添加任何终结点,则运行时添加默认终结点。

(11) 为了确保正确承载该服务,请打开 Internet Explorer 的实例,导航到该服务的 URL:http://localhost/IISHostedCalc/Service.svc。

2. 示例

下面是 IIS 承载的计算器服务的代码的完整列表。

```csharp
using System;
using System.ServiceModel;
namespace Microsoft.ServiceModel.Samples
{
  [ServiceContract]
  public interface ICalculator
  {
    [OperationContract]
    double Add(double n1, double n2);
    [OperationContract]
    double Subtract(double n1, double n2);
    [OperationContract]
    double Multiply(double n1, double n2);
    [OperationContract]
    double Divide(double n1, double n2);
  }
```

```csharp
public class CalculatorService : ICalculator
{
    public double Add(double n1, double n2)
    {
        return n1 + n2;
    }
    public double Subtract(double n1, double n2)
    {
        return n1 - n2;
    }
    public double Multiply(double n1, double n2)
    {
        return n1 * n2;
    }
    public double Divide(double n1, double n2)
    {
        return n1 / n2;
    }
}
```

XML 文件:

```xml
<?xml version="1.0" encoding="utf-8"?>
<configuration>
  <system.serviceModel>
    <services>
      <service name="Microsoft.ServiceModel.Samples.CalculatorService">
        <!-- This endpoint is exposed at the base address provided by host: http://localhost/servicemodelsamples/service.svc -->
        <endpoint address=""
                  binding="wsHttpBinding"
                  contract="Microsoft.ServiceModel.Samples.ICalculator" />
        <!-- The mex endpoint is explosed at http://localhost/servicemodelsamples/service.svc/mex -->
        <endpoint address="mex"
                  binding="mexHttpBinding"
                  contract="IMetadataExchange" />
      </service>
    </services>
```

```
 </system.serviceModel>
</configuration>
```

7.2 WAS 承载

  Windows 进程激活服务（WAS）管理辅助进程的激活和生存期，该辅助进程包含承载 Windows Communication Foundation（WCF）服务的应用程序。WAS 进程模型通过移除对 HTTP 的依赖性使 HTTP 服务器的 IIS 6.0 进程模型通用化。这将允许 WCF 服务在宿主环境中同时使用 HTTP 和非 HTTP 协议（如 Net.TCP），该宿主环境支持基于消息的激活并提供在给定计算机上承载大量应用程序的能力。

  WAS 进程模型提供了一些功能，可以以一种更为可靠、更易管理并有效地使用资源的方式承载应用程序：

  (1) 基于消息的应用程序激活和辅助进程应用程序会动态地启动和停止，以响应使用 HTTP 和非 HTTP 网络协议送达的传入工作项。

  (2) 可靠的应用程序和辅助进程回收可以使应用程序保持良好的运行状况。

  (3) 集中的应用程序配置和管理。

  (4) 允许应用程序利用 IIS 进程模型，而无需完全 IIS 安装的部署需求量。

  Windows Server AppFabric 配合 IIS 7.0 和 Windows 进程激活服务（WAS）提供丰富的应用程序宿主环境为 NET4 WCF 和 WF 服务。这些优点包括进程生命周期管理、进程回收、共享承载、快速失败保护、进程孤立、按需激活和运行状况监视。

1. WAS 寻址模型的元素

  应用程序具有统一资源标识符（URI）地址，这些地址是一些代码单元，其生存期和执行环境由服务器管理。一个 WAS 服务器实例可以承载多个不同的应用程序。服务器将分组称为应用程序组织站点。在网站中，应用程序是以分层的方式排列的，这种方式反映了充当其外部地址的 URI 的结构。

  应用程序地址分为两个部分：基本 URI 前缀和应用程序特定的相对地址（路径）。这两个部分结合在一起时可提供应用程序的外部地址。基本 URI 前缀从网站绑定构造的，并且适用于网站中的所有应用程序。然后通过以下方式构造应用程序地址：采用应用程序特定的路径片段（如"/applicationOne"）并将其追加到基 URI 前缀（例如"net.tcp://localhost"），以形成完整的应用程序 URI。

  表 7-1 列举了使用 HTTP 和非 HTTP 网站绑定的 WAS 网站的几个可能的寻址方案。

表7-1 可能的寻址方案

方案	网站绑定	应用程序路径	基应用程序 URI
仅 HTTP	http：*：80：*	/appTwo	http://localhost/appTwo/
HTTP 和非 HTTP	http：*：80：* net.tcp：808：*	/appTwo	http://localhost/appTwo/ net.tcp://localhost/appTwo/
仅非 HTTP	net.pipe：*	/appThree	net.pipe://appThree/

也可以对应用程序内的服务和资源进行寻址。在应用程序内,根据其自身的路径对应用程序资源进行寻址。例如,假定计算机上名为 contoso.com 的网站同时具有 HTTP 和 Net.TCP 协议的网站绑定。还假定该网站包含一个位于 /Billing 处的应用程序,该应用程序在 GetOrders.svc 中公开服务。然后,如果 GetOrders.svc 服务使用 SecureEndpoint 的相对地址公开了一个终结点,则将会在下面的两个 URI 中公开该服务终结点:

http://contoso.com/Billing/GetOrders.svc/SecureEndpoint
net.tcp://contoso.com/Billing/GetOrders.svc/SecureEndpoint

2. WAS 运行库

为了便于寻址和管理,可将应用程序组织到网站中。运行时,还会将应用程序分组到应用程序池中。一个应用程序池可以存放多个来自许多不同网站的应用程序。一个应用程序池中的所有应用程序共享一组公共的运行时特征。例如,它们都在公共语言运行库(CLR)的同一版本下运行,并都共享一个公共的进程标识。每个应用程序池都与辅助进程(w3wp.exe)的某个实例相对应。通过 CLR AppDomain,在共享应用程序池内运行的每个托管应用程序都独立于其他的应用程序。

7.2.1 WAS 激活体系结构

1. 激活组件

WAS 由几个体系结构组件组成:

(1)侦听器适配器。通过特定的网络协议接收消息并与 WAS 进行通信以将传入消息路由到正确的辅助进程中的 Windows 服务。

(2)WAS。管理工作进程的创建和生存期的 Windows 服务。

(3)一般辅助进程可执行程序(w3wp.exe)。见图 7-1。

(4)应用程序管理器。管理在辅助进程中承载应用程序的应用程序域的创建和生存期。

（5）协议处理程序。在辅助进程中运行并管理辅助进程与各个侦听器适配器之间的通信协议组件。存在两种类型的协议处理程序：进程协议处理程序和 AppDomain 协议处理程序。

当 WAS 激活辅助进程实例时，会将所需的进程协议处理程序加载到辅助进程中，并使用应用程序管理器来创建一个应用程序域以承载应用程序。应用程序域将加载应用程序的代码以及应用程序使用的网络协议所要求的 AppDomain 协议处理程序。

图 7-1　一般辅助进程 w3wp.exe

2. 侦听器适配器

侦听器适配器是一些单独的 Windows 服务，这些服务可以实现用于通过其侦听的网络协议接收消息的网络通信逻辑。表 7-2 列出了 Windows Communication Foundation（WCF）协议的侦听器适配器。

表 7-2　WCF 协议的侦听器适配器

侦听器适配器服务名称	协议	说明
W3SVC	http	为 IIS 7.0 和 WCF 提供 HTTP 激活的公共组件
NetTcpActivator	net.tcp	取决于 NetTcpPortSharing 服务
NetPipeActivator	net.pipe	
NetMsmqActivator	net.msmq	适用于基于 WCF 的消息队列应用程序
NetMsmqActivator	msmq.formatname	提供与现有消息队列应用程序的向后兼容性

在安装过程中，在 applicationHost.config 文件中注册特定协议的侦听器适配器，如下面的 XML 示例中所示。

`<system.applicationHost>`

```
<listenerAdapters>
    <add name = "http" />
    <add name = "net.tcp"
        identity = "S-1-5-80-3579033775-2824656752-1522793541-1960352512-462907086" />
     <add name = "net.pipe"
        identity = "S-1-5-80-2943419899-937267781-4189664001-1229628381-3982115073" />
      <add name = "net.msmq"
        identity = "S-1-5-80-89244771-1762554971-1007993102-348796144-2203111529" />
        <add name = "msmq.formatname"
          identity = "S-1-5-80-89244771-1762554971-1007993102-348796144-2203111529" />
</listenerAdapters>
</system.applicationHost>
```

3. 协议处理程序

在计算机级别的 Web.config 文件中注册特定协议的进程和 AppDomain 协议处理程序。

```
<system.web>
  <protocols>
    <add name = "net.tcp"
     processHandlerType =
      "System.ServiceModel.WasHosting.TcpProcessProtocolHandler"
     appDomainHandlerType =
      "System.ServiceModel.WasHosting.TcpAppDomainProtocolHandler"
     validate = "false" />
    <add name = "net.pipe"
     processHandlerType =
      "System.ServiceModel.WasHosting.NamedPipeProcessProtocolHandler"
       appDomainHandlerType =
      "System.ServiceModel.WasHosting.NamedPipeAppDomainProtocolHandler" />
    <add name = "net.msmq"
     processHandlerType =
      "System.ServiceModel.WasHosting.MsmqProcessProtocolHandler"
     appDomainHandlerType =
      "System.ServiceModel.WasHosting.MsmqAppDomainProtocolHandler"
```

```
            validate = "false" />
    </protocols>
</system.web>
```

7.2.2 WAS 配置

本节介绍在 Windows Vista 中设置 Windows 进程激活服务(也称为 WAS)使其承载不通过 HTTP 网络协议进行通信的 Windows Communication Foundation (WCF) 服务所需的步骤。下面的部分略述此配置的步骤:

◇ 安装(或确认安装)所需的 WCF 激活组件。
◇ 创建一个具有要使用的网络协议绑定的 WAS 站点,或者向现有站点添加新协议绑定。
◇ 创建一个应用程序以承载服务,并使该应用程序可以使用所需的网络协议。
◇ 生成一个公开非 HTTP 终结点的 WCF 服务。

1. 使用非 HTTP 绑定配置站点

若要将非 HTTP 绑定与 WAS 一起使用,必须将站点绑定添加到 WAS 配置。WAS 的配置存储是 applicationHost.config 文件,该文件位于 %windir%\system32\inetsrv\config 目录中。此配置存储由 WAS 和 IIS 7.0 共享。

applicationHost.config 是一个 XML 文本文件,可以使用任何标准文本编辑器(如记事本)打开。不过,IIS 7.0 命令行配置工具(appcmd.exe)是添加非 HTTP 站点绑定的首选方法。

下面的命令使用 appcmd.exe 将 net.tcp 站点绑定添加到默认网站(将此命令作为单独的一行输入)。

```
appcmd.exe set site "Default Web Site" -+bindings.[protocol='net.tcp',bindingInformation='808:*']
```

通过将下面指示的行添加到 applicationHost.config 文件,此命令将新 net.tcp 绑定添加到默认网站。

```
<sites>
    <site name = "Default Web Site" id = "1">
        <bindings>
            <binding protocol = "HTTP" bindingInformation = "*:80:" />
            //The following line is added by the command.
            <binding protocol = "net.tcp" bindingInformation = "808:*" />
        </bindings>
    </site>
```

< /sites >

2. 使应用程序可以使用非 HTTP 协议

可以启用或禁用单个网络 protocolsat 应用程序级别。下面的命令说明如何为在 Default Web Site 中运行的应用程序同时启用 HTTP 和 net.tcp 协议。

```
appcmd.exe set app "Default Web Site/appOne" /enabledProtocols:net.tcp
```

也可以在设置启用的协议列表 < applicationDefaults > 中存储 ApplicationHost.config 的 XML 配置元素。

摘自 applicationHost.config 的以下 XML 代码说明一个已同时绑定到 HTTP 协议和非 HTTP 协议的站点。支持非 HTTP 协议所需的其他配置通过注释进行了突出。

```
< sites >
    < site name = "Default Web Site" id = "1" >
      < application path = "/" >
         < virtualDirectory path = "/" physicalPath = "D:\inetpub\wwwroot" />
      < /application >
        < bindings >
           //The following two lines are added by the command.
           < binding protocol = "HTTP" bindingInformation = " * :80:" />
           < binding protocol = "net.tcp" bindingInformation = "808: * " />
        < /bindings >
    < /site >
    < siteDefaults >
       < logFile
       customLogPluginClsid = "{FF160663 - DE82 - 11CF - BC0A - 00AA006111E0}"
         directory = "D:\inetpub\logs\LogFiles" />
       < traceFailedRequestsLogging
         directory = "D:\inetpub\logs\FailedReqLogFiles" />
    < /siteDefaults >
    < applicationDefaults
      applicationPool = "DefaultAppPool"
      //The following line is inserted by the command.
      enabledProtocols = "http, net.tcp" />
    < virtualDirectoryDefaults allowSubDirConfig = "true" />
< /sites >
```

如果用户尝试通过用于非 HTTP 激活的 WAS 来激活服务,并且用户未安装和配置 WAS,可能会看到以下错误:

[InvalidOperationException: The protocol 'net.tcp' does not have an implementation of HostedTransportConfiguration type registered.] System. ServiceModel. AsyncResult. End (IAsyncResult result) + 15778592 System. ServiceModel.Activation.HostedHttpRequestAsyncResult.End(IAsyncResult result) + 15698937 System. ServiceModel. Activation. HostedHttpRequestAsyncResult. ExecuteSynchronous(HttpApplication context, Boolean flowContext) +265 System. ServiceModel.Activation.HttpModule.ProcessRequest(Object sender, EventArgs e) + 227 System. Web. SyncEventExecutionStep. System. Web. HttpApplication. IExecutionStep. Execute ( ) + 80 System. Web. HttpApplication. ExecuteStep (IExecutionStep step, Boolean& completedSynchronously) +171

如果看到此错误,确保已安装并正确配置了用于非 HTTP 激活的 WAS。

3. 生成一个将 WAS 用于非 HTTP 激活的 WCF 服务

执行安装和配置 WAS 的步骤后(如前所述),将服务配置为使用 WAS 进行激活与配置承载于 IIS 中的服务类似。

7.2.3　WAS 承载服务方法

本节概述了创建由 Windows 进程激活服务(也称为 WAS)承载的 Windows Communication Foundation (WCF)服务所需的基本步骤。WAS 是新的进程激活服务,是对使用非 HTTP 传输协议的 Internet Information Services (IIS) 功能的泛化。WCF 使用监听器适配器接口传递激活请求,这些请求是通过由 WCF 支持的非 HTTP 协议(如 TCP、命名管道和消息队列)收到的。

此主机选项要求正确安装和配置 WAS 激活组件,但不要求编写任何主机代码作为应用程序的一部分。

警告

如果将 Web 服务器的请求处理管道设置为经典模式,则将不支持 WAS 激活。如果要使用 WAS 激活,则必须将 Web 服务器的请求处理管道设置为集成模式。

在 WAS 中承载 WCF 服务时,可按照通常的方式使用标准绑定。但是,在使用 NetTcpBinding 和 NetNamedPipeBinding 配置 WAS 承载的服务时,必须满足一个约束条件。当不同的终结点使用相同的传输时,绑定设置必须在以下的 7 个属性上相匹配:

- ConnectionBufferSize
- ChannelInitializationTimeout
- MaxPendingConnections
- MaxOutputDelay
- MaxPendingAccepts
- ConnectionPoolSettings. IdleTimeout

◇ ConnectionPoolSettings. MaxOutboundConnectionsPerEndpoint

否则,最先初始化的终结点总是确定这些属性的值,并且以后添加的终结点若与这些设置不匹配,则将引发 ServiceActivationException。

1. 创建 WAS 承载的基本服务

(1)为该类型的服务定义服务协定。

```
[ServiceContract]
public interface ICalculator
{
   [OperationContract]
   double Add(double n1, double n2);
   [OperationContract]
   double Subtract(double n1, double n2);
   [OperationContract]
   double Multiply(double n1, double n2);
   [OperationContract]
   double Divide(double n1, double n2);
}
```

(2)在服务类中实现该服务协定。请注意,在服务的实现内部,未指定地址或绑定信息。而且,不必编写代码也可从配置文件中检索该信息。

```
public class CalculatorService : ICalculator
{
   public double Add(double n1, double n2)
   {
      return n1 + n2;
   }
   public double Subtract(double n1, double n2)
   {
      return n1 - n2;
   }
   public double Multiply(double n1, double n2)
   {
      return n1 * n2;
   }
   public double Divide(double n1, double n2)
   {
      return n1 /n2;
```

}
}
```

(3) 创建 Web.config 文件,以定义要由 NetTcpBinding 终结点使用的 CalculatorService 绑定。

```xml
<?xml version="1.0" encoding="utf-8"?>
<configuration>
  <system.serviceModel>
    <bindings>
      <netTcpBinding>
        <binding portSharingEnabled="true">
          <security mode="None"/>
        </binding>
      </netTcpBinding>
    </bindings>
  </system.serviceModel>
</configuration>
```

(4) 创建包含以下代码的 Service.svc 文件。

```
<%@ ServiceHost language=c# Service="CalculatorService" %>
```

(5) 将 Service.svc 文件放到 IIS 虚拟目录中。

2. 创建要使用服务的客户端

(1) 使用 ServiceModel 元数据实用工具(Svcutil.exe)从命令行根据服务元数据生成代码。

```
Svcutil.exe <service's Metadata Exchange (MEX) address or HTTP GET address>
```

(2) 生成的客户端包含 ICalculator 接口,该接口定义了客户端实现必须满足的服务协定。

```
[System.ServiceModel.ServiceContractAttribute(
Namespace="http://Microsoft.ServiceModel.Samples", ConfigurationName="Microsoft.ServiceModel.Samples.ICalculator")]
public interface ICalculator
{ [System.ServiceModel.OperationContractAttribute(
Action="http://Microsoft.ServiceModel.Samples/ICalculator/Add", ReplyAction="http://Microsoft.ServiceModel.Samples/ICalculator/AddResponse")]
    double Add(double n1, double n2);
    [System.ServiceModel.OperationContractAttribute(
Action="http://Microsoft.ServiceModel.Samples/ICalculator/Subtract", ReplyAction="http://Microsoft.ServiceModel.Samples/ICalculator/
```
```

```
SubtractResponse")]
        double Subtract(double n1, double n2);
            [System.ServiceModel.OperationContractAttribute(
    Action = " http://Microsoft.ServiceModel.Samples/ICalculator/Multiply ",
ReplyAction = " http://Microsoft.ServiceModel.Samples/ICalculator/
MultiplyResponse")]
        double Multiply(double n1, double n2);

            [System.ServiceModel.OperationContractAttribute(
    Action = " http://Microsoft.ServiceModel.Samples/ICalculator/Divide ",
ReplyAction = " http://Microsoft.ServiceModel.Samples/ICalculator/
DivideResponse")]
        double Divide(double n1, double n2);
}
```

(3) 生成的客户端应用程序还包含 ClientCalculator 的实现。请注意,在服务的实现内部,未指定地址和绑定信息。而且,不必编写代码也可从配置文件中检索该信息。

```
    public partial class CalculatorClient : System.ServiceModel.ClientBase <
Microsoft.ServiceModel.Samples.ICalculator >, Microsoft.ServiceModel.
Samples.ICalculator
{ public CalculatorClient()
    { }
        public CalculatorClient(string endpointConfigurationName) :
            base(endpointConfigurationName)
    { }
        public CalculatorClient ( string endpointConfigurationName, string
remoteAddress) :
            base(endpointConfigurationName, remoteAddress)
    { }
        public CalculatorClient(string endpointConfigurationName,
    System.ServiceModel.EndpointAddress remoteAddress) :
            base(endpointConfigurationName, remoteAddress)
    { }
        public CalculatorClient(System.ServiceModel.Channels.Binding binding,
System.ServiceModel.EndpointAddress remoteAddress) :
            base(binding, remoteAddress)
    { }
        public double Add(double n1, double n2)
```

```
            return base.Channel.Add(n1, n2);    }
        public double Subtract(double n1, double n2)
    {       return base.Channel.Subtract(n1, n2);    }
        public double Multiply(double n1, double n2)
    {       return base.Channel.Multiply(n1, n2);    }
        public double Divide(double n1, double n2)
    {       return base.Channel.Divide(n1, n2);    }}
```

(4) 使用 NetTcpBinding 的客户端配置也通过 Svcutil.exe 生成。在使用 Visual Studio 时,应在 App.config 文件中命名此文件。

```
<?xml version = "1.0" encoding = "utf-8"?>
<configuration>
    <system.serviceModel>
        <bindings>
            <netTcpBinding>
                <binding name = "NetTcpBinding_ICalculator">
                    <security mode = "None"/>
                </binding>
            </netTcpBinding>
        </bindings>
        <client>
            <endpoint
              address = "net.tcp://localhost/servicemodelsamples/service.svc"
                binding = "netTcpBinding" bindingConfiguration = "NetTcpBinding_ICalculator"
                contract = "ICalculator" name = "NetTcpBinding_ICalculator"/>
        </client>
    </system.serviceModel>
</configuration>
```

(5) 在应用程序中创建 ClientCalculator 的实例,然后调用服务操作。

```
class Client{
    static void Main()
    {  CalculatorClient client = new CalculatorClient();
        double value1 = 100.00D;
        double value2 = 15.99D;
        double result = client.Add(value1, value2);
        Console.WriteLine("Add({0},{1}) = {2}", value1, value2, result);
        value1 = 145.00D;   value2 = 76.54D;
```

```
result = client.Subtract(value1, value2);
Console.WriteLine("Subtract({0},{1}) = {2}", value1, value2, result);
//Call the Multiply service operation.
value1 = 9.00D; value2 = 81.25D;
result = client.Multiply(value1, value2);
Console.WriteLine("Multiply({0},{1}) = {2}", value1, value2, result);
//Call the Divide service operation.
value1 = 22.00D; value2 = 7.00D;
result = client.Divide(value1, value2);
Console.WriteLine("Divide({0},{1}) = {2}", value1, value2, result);
client.Close();
Console.WriteLine();
Console.WriteLine("Press <ENTER> to terminate client.");
Console.ReadLine();    }  }
```

(6) 编译并运行客户端。

7.3 Windows 服务应用程序中承载

7.3.1 概述

Windows 服务(以前称为 Windows NT 服务)提供了一种尤其适合于下面这样的应用程序的进程模型：必须在长时间运行的可执行程序中生存，并且不显示任何形式的用户界面。Windows 服务应用程序的进程生存期由服务控制管理器（SCM）管理，可以通过该管理器启动、停止和暂停 Windows 服务应用程序。可以配置 Windows 服务进程启动计算机，使其在成为"始终运行的"应用程序的合适的宿主环境时自动启动。

承载长时间运行的 Windows Communication Foundation（WCF）服务的应用程序具有许多和 Windows 服务一样的特性。具体而言，WCF 服务是长时间运行的服务器可执行程序，并且不与用户直接交互，因此也不实现任何形式的用户界面。因此，在 Windows 服务应用程序内承载 WCF 服务是一个用于生成可靠的、长时间运行的 WCF 应用程序的可行方案。

通常，WCF 开发人员必须决定是在 Windows 服务应用程序内、Internet 信息服务（IIS）内还是在 Windows 进程激活服务（WAS）宿主环境中承载他们的 WCF 应用程序。在下列条件下，应考虑使用 Windows 服务应用程序：

◇ 应用程序要求显式激活。例如，在服务器启动时应用程序必须自动启动（而不是动态启动以响应第一个传入消息）的条件下，应使用 Windows 服务。

◇ 承载应用程序的进程必须在启动后保持运行状态。Windows 服务进程一旦启动将一直保持运行状态,直到服务器管理员使用服务控制管理器显式关闭该进程。在 IIS 或 WAS 中承载的应用程序可能会动态地启动和停止,以便最佳地使用系统资源。需要显式控制其宿主进程的生存期的应用程序应使用 Windows 服务而非 IIS 或 WAS。

◇ WCF 服务必须在 Windows Server 2003 上运行并且必须使用 HTTP 之外的传输。在 Windows Server 2003 上,IIS 6.0 宿主环境限制为仅可进行 HTTP 通信。Windows 服务应用程序不受此限制的约束,可以使用 WCF 支持的任何传输协议,包括 net.tcp、net.pipe 和 net.msmq。

1. 在 Windows 服务应用程序内承载 WCF

(1)创建 Windows 服务应用程序。可以使用 System.ServiceProcess 命名空间中的类以托管代码的形式编写 Windows 服务应用程序。此应用程序必须包含一个继承自 ServiceBase 的类。

(2)将 WCF 服务的生存期链接到 Windows 服务应用程序的生存期。通常,人们希望 Windows 服务应用程序中承载的 WCF 服务能够在宿主服务启动时变为活动状态,而在宿主服务停止时停止对消息的侦听,并在 WCF 服务遇到错误时关闭宿主进程。这可以通过以下操作实现:

◇ 重写 OnStart(String[ ]) 以打开一个或多个 ServiceHost 实例。一个 Windows 服务应用程序可以承载多个作为一个组同时启动和停止的 WCF 服务。

◇ 重写 OnStop 以在 Closed 任何运行的 ServiceHost 服务上调用 WCF,这些服务在 OnStart(String[ ])过程中启动。

◇ 订阅 Faulted 的 ServiceHost 事件,并使用 ServiceController 类以在出现错误时关闭 Windows 服务应用程序。

部署和管理承载 WCF 服务的 Windows 服务应用程序的方式与不使用 WCF 的 Windows 服务应用程序的一样。

7.3.2 托管 Windows 服务中承载 WCF 服务

本节概述了创建由 Windows 服务承载的 Windows Communication Foundation(WCF)服务所需的基本步骤。此方案可通过托管 Windows 服务承载选项启用,此选项是在没有消息激活的安全环境中在 Internet 信息服务(IIS)外部承载的、长时间运行的 WCF 服务。服务的生存期改由操作系统控制。此宿主选项在 Windows 的所有版本中都是可用的。

可以使用 Microsoft 管理控制台(MMC)中的 Microsoft.ManagementConsole.SnapIn 管理 Windows 服务,并且可以将其配置为在系统启动时自动启动。此承载选项包括注册承载

WCF 服务作为托管 Windows 服务的应用程序域,因此服务的进程生存期由 Windows 服务的服务控制管理器(SCM)来控制。

服务代码包括服务协定的服务实现、Windows 服务类和安装程序类。服务实现类 CalculatorService 是 WCF 服务。CalculatorWindowsService 是 Windows 服务。要符合 Windows 服务的要求,该类继承自 ServiceBase 并实现 OnStart 和 OnStop 方法。在 OnStart 中,将为 ServiceHost 类型创建 CalculatorService 并打开它。在 OnStop 中,停止并释放服务。主机还负责提供服务主机基址,该基址已在应用程序设置中进行设置。安装程序类继承自 Installer,允许程序通过 Installutil.exe 工具安装为 Windows 服务。

1. 构造服务并提供宿主代码

(1)使用 Service.cs 文件中的计算器服务接口来定义 ICalculator 服务协定。

(2)通过从 WCF ICalculator 接口继承,在 Service.cs 文件的 CalculatorService 类中实现服务协定作为 WCF 服务。

(3)通过从 ServiceBase 类继承来实现 Windows 服务。重写 OnStart 方法以创建并打开 ServiceHost 的实例。重写 OnStop 方法以关闭 ServiceHost 的实例。创建 CalculatorService 实例的实例并将其命名为"WCFWindowsServiceSample"。提供应用程序的入口点。

(4)创建 ProjectInstaller 类,此类继承自 Installer 并用设置为 true 的 RunInstallerAttribute 来标记,因此,在安装程序集时将调用 Visual Studio 的自定义操作安装程序或 Installutil.exe。

(5)在配置中提供服务的基址。

2. 安装并运行服务

(1)编译服务以生成 Service.exe 可执行文件。

(2)在命令提示符处键入 installutil bin\service.exe 来安装 Windows 服务。(如果尚未设置该工具的路径,则该工具放置在 Microsoft.NET Framework 安装目录中。)在命令提示符处键入 services.msc 以访问服务控制管理器(SCM)。WINDOWS 服务应作为"WCFWindowsServiceSample"出现在服务中。只有在 Windows 服务正在运行的情况下,WCF 服务才能响应客户端。若要启动服务,请在 SCM 中右击服务并选中"启动",或在命令提示符处键入 net start WCFWindowsServiceSample。

(3)如果对服务进行更改,则必须首先停止并卸载服务。若要停止服务,请在 SCM 中右击服务并选中"停止",或在命令提示符处键入 net stop WCFWindowsServiceSample。请注意,如果停止 WINDOWS 服务然后运行客户端,则在客户端尝试访问服务时发生 EndpointNotFoundException 异常。若要卸载 WINDOWS 服务,请在命令提示符处键入 installutil /u bin\service.exe。

7.4 托管应用程序中承载

Windows Communication Foundation（WCF）服务可以承载于任何 .NET Framework 应用程序中。自承载服务是最灵活的宿主选项，因为此服务部署所需要的基础结构最少。但是，此服务也是最不可靠的宿主选项，因为托管应用程序未提供 WCF 中其他宿主选项（如 Internet 信息服务（IIS）和 Windows 服务）的高级宿主和管理功能。

若要创建自承载服务，请创建并打开 ServiceHost 的实例以启动侦听消息的服务。以下各部分描述了使用此宿主选项的常见情况。

1. 控制台应用程序

自承载支持的常见方案是在控制台应用程序内部运行的 WCF 服务。在控制台应用程序内部承载一个 WCF 服务通常在服务的开发阶段非常有用。这使服务变得容易调试，从中跟踪信息以查明应用程序内发生的情况变得更加方便，以及通过将其复制到新的位置进行来回移动变得更加轻松。

2. 胖客户端应用程序

自承载支持的其他常见方案是胖客户端应用程序，如基于 Windows Presentation Foundation（WPF）或 Windows 窗体（WinForms）的应用程序。此宿主选项还使丰富客户端应用程序（如 WPF 和 WinForms 应用程序）与外部世界的通信变得很容易。例如，一个将 WPF 用于其用户界面并作为 WCF 服务主机的对等协作客户端，允许其他客户端连接到它并共享信息。

参 考 文 献

[1] JUVAL LOWY. WCF 编程[M]. 张逸,徐宁, 译. 北京:机械工业出版社,2008.
[2] 马骏. C#网络应用编程[M]. 3 版. 北京:人民邮电出版社,2014.
[3] 张敬普,丁士锋. C#5.0 与.NET4.5 高级编程 – LINQ、WCF、WPF 和 WF[M]. 北京:清华大学出版社,2014.
[4] 周家安. WCF 编程权威指南[M]. 北京:清华大学出版社,2018.